PHYSICS OF SPORTS

SELECTED REPRINTS

Edited by

Cliff Frohlich

published by

American Association of Physics Teachers

Published by:
American Association of Physics Teachers
Publications Department
5110 Roanoke Place, Suite 101
College Park, MD 20740, U.S.A.

Cover Art: From "The Influence of Track Compliance on Running," by T.A. McMahon and P.R. Greene. *Journal of Biomechanics,* **12**, p. 898, fig. 4. Copyright: Pergamon Press Ltd. See article beginning on p. 93.

ISBN #0-917853-24-5

Contents

RESOURCE LETTER

Roger H. Stuewer, *Editor*

School of Physics and Astronomy, 116 Church Street
University of Minnesota, Minneapolis, Minnesota 55455

This is one of a series of Resource Letters on different topics intended to guide college physicists, astronomers, and other scientists to some of the literature and other teaching aids that may help improve course content in specified fields. No Resource Letter is meant to be exhaustive and complete; in time there may be more than one letter on some of the main subjects of interest. Comments on these materials as well as suggestions for future topics will be welcomed. Please send such communications to Professor Roger H. Stuewer, Editor, AAPT Resource Letters, School of Physics and Astronomy, 116 Church Street SE, University of Minnesota, Minneapolis, MN 55455.

Resource letter PS-1: Physics of sports

Cliff Frohlich
Institute for Geophysics, University of Texas at Austin, Austin, Texas 78713

(Received 11 December 1985; accepted for publication 20 December 1985)

This Resource Letter provides a guide to the literature on the physics of sports. The letter E after an item indicates elementary level or material of general interest to persons becoming informed in the field. The letter I, for intermediate level, indicates material of a somewhat more specialized nature; and the letter A indicates rather specialized or advanced material. An asterisk (*) indicates those articles to be included in an accompanying Reprint Book.

I. INTRODUCTION

Human interest in sports is so widespread that a substantial fraction of the world's population follows progress in events such as the Olympic Games or the World Cup soccer matches. Although sport is neither life threatening nor economically important to most people, about one-quarter of most daily newscasts specifically concerns sports. We can safely presume that a substantial fraction of physicists are or have been sports enthusiasts, fans, or participants.

Nevertheless, there is surprisingly little published information about the basic physics underlying most sports, even though the relevant physics is all classical. While there are a few excellent books about the physics of sports, none contain an exhaustive set of references. There are research papers published in biomechanics or physical education journals about sports; however, the focus of much of this literature differs from the focus of the physicist.

We can identify three reasons why the literature concerning the physics of sports is so sparse. One reason is because sports as we know them today are relatively recent phenomena. Most of the sports played today, such as basketball, baseball, football, and soccer, are only about a century old, and their association with academic centers such as high schools and colleges has occurred only since about World War I. The highly technical approach to training, etc., for professional and Olympic sports developed only since World War II. A second reason is that sports have only recently been perceived as a legitimate topic for scientific analysis. Just as human psychology and sexual behavior were not "respectable" topics for scientific research a century ago, until recently sports have been perceived as little more than trivial amusements and games. Finally, many scientists wrongly assume that the physics of particular sporting events must have been worked out explicitly somewhere, even though they know it does not appear in the *Physical Review*.

In individual sporting events, the depth of research analysis varies widely depending on social and economic factors, as well as on the specific rules of particular sports. For example, there were papers published about the physics of golf in the 19th century, when both golf and classical mechanics were apparently respectable pursuits for physicists. More recently research has continued because the rules allow the golfer to choose both the ball and his or her clubs, creating a rather lucrative market for the physicist, engineer, or crackpot who can design equipment which is, or apparently is, more effective. In contrast to golf, baseball has developed more recently, and there are fewer socioeconomic pressures on equipment design. Moreover, the rules specifically restrict the dimensions and material character of the bat. The umpire, and not the players, chooses the ball. Consequently there has been less research about baseballs and bats than about golf equipment.

The rules in some sports discourage research. For example, while the basic design of the present bicycle developed before 1900 because of interest in racing, its upright, diamond shaped frame was fixed by the rules in 1938 after more aerodynamically sound reclining bicycles began winning races. In track and field, experiments in wind tunnels produced more effective javelins for a few fortunate athletes, until the shape of the javelin was fixed by rules committees in 1961. Although the discus is influenced as greatly as the javelin by aerodynamic forces, it has never enjoyed the concentrated attention of researchers because its shape was fixed when rules were first put forth near the beginning of this century.

Investigators of several different types regularly or occasionally undertake research concerning sporting events or equipment:

(1) Researchers in *physical education* are probably the most sincerely interested in sport itself. Their research efforts tend to be concentrated in the practical aspects of training and performance, and articles in physical education journals often incorporate little or no physics in their analyses. However, because these researchers are often athletes or have interacted regularly with athletes, they often have insight which allows them to predict the results of physical experiments.

(2) Researchers in *biomechanics* traditionally study subjects such as the mechanics of joints, arteries, bone fracture, and also "normal" activities such as walking, etc. A typical biomechanical analysis of an athletic activity might start by filming the activity with a high-speed camera, and then digitizing the position of the athlete's joints on the film, frame by frame. These studies usually can provide a physicist with the best available information as to what actually happens in an athletic activity. Perhaps because of its association with medicine, the field of biomechanics is well organized, and has several journals and professional organizations.

(3) Probably the most thorough and sophisticated research concerning the physics of sports has been performed by *mechanical engineers*. However, these studies appear only rarely in journal articles, and the author is unaware of any organization of sports engineers. Many of the studies are commissioned by equipment companies and other private organizations. The report of these studies are often proprietary.

(4) Finally, a few *physicists* dabble in sports topics, but at present there are no journals, funding agencies, or professional organizations which specifically encourage research in the physics of sports. Many, but not all, of the available journal articles are concerned with sports topics solely as examples for teaching physics, and thus do not report careful measurements of, or attempt to propose realistic models for, sporting activity.

One difficulty facing the scientist who might undertake sports physics research is that although sport science is an interdisciplinary field, most journals have a narrow perspective. Suppose, for example, that you undertook a research program on the factors affecting hitting in baseball. Suppose further that your research was successful, and you wrote a comprehensive paper describing your results, including the physics of the air/ball and bat/ball interaction, the physiology of visual perception of the ball, the mechanics of the swing, the results from analyzing many game situations, and some practical implications for pitching and hitting strategy. Most likely, *no journal would publish your entire paper*. *American Journal of Physics* would publish the equations describing your model of bat, ball, and swing; *International Journal of Sport Biomechanics* would publish the analysis of films of batters; *Vision Research* would publish the evaluation of visual perception of batters; *Research Quarterly* would publish suggestions for improving the training of batters; and *Baseball Research Journal* would publish the historical data of real games. Conversely, at present the journal articles containing the best physics analysis are usually not those having the best measurements or the most quantitative evaluation of strategy.

Nevertheless, one attraction of sports physics research is that one can produce publishable research by performing relatively inexpensive experiments, or by undertaking relatively simple calculations. One can design these research projects so that students can participate and use equipment that is generally available in most college labs. For example, there is surprisingly little basic data concerning aerodynamic effects on balls. When maximum velocities are reported for baseball pitches or tennis serves, the reports usually ignore the possible effects of air resistance and do not distinguish whether the velocity is for the beginning or the end of the trajectory. While encyclopedias commonly report that jai-alai is the "fastest game" and that balls attain speeds of 150 miles/h, I have been unable to find how this measurement was made, or who made it. The available measurements of baseball/bat contact time are for balls hit from a tee, determined by stroboscopic techniques. However, one could perform a better measurement using a sensor attached to the bat. Possibly one could even market such a device to baseball and golf enthusiasts if it directly measured momentum transfer. In spite of several papers on the subject, there is still confusion as to whether the rebound velocity of tennis balls differs for clamped and unclamped rackets. I am unaware of any experimental or theoretical treatment describing what happens to the air pressure or leather cover of a football or soccer ball as it is kicked. Indeed, there exist remarkably few published discussions of the physics of football or soccer at all.

II. JOURNALS

American Journal of Physics—This journal publishes a few sports physics papers each year because of their relevance to the "instructional and cultural aspects of physical science." Although the papers often are deficient in referencing previous work in physical education or biomechanics, they usually show more concern for the essential concepts of physics in sports than do papers in other journals.

International Journal of Sport Biomechanics—This is a new journal (first issue, February 1985) concentrating specifically on the mechanical analysis of athletes and sporting events.

Journal of Applied Mechanics—This journal rarely publishes sports papers, but these few papers typically embody a sophisticated classical engineering mechanics analysis.

Journal of Biomechanics—Papers mostly concentrate on the mechanics of joints, bone fracture, arterial flow, and other quasimedical topics, but occasionally there are papers analyzing the physics in specific sports.

Medicine and Science in Sports and Exercise—This journal concentrates on the physiological and medical effects of sport and exercise; however, it occasionally publishes papers about sport biomechanics.

Research Quarterly for Exercise and Sport—This journal seldom if ever publishes sports physics papers in a strict sense, but quite commonly reports physical education research concerning ball speed, bodily characteristics of athletes during performance, etc. For the physicist it can be a valuable source of data.

III. BOOKS

1. **Patterns of Human Motion—A Cinematographic Analysis**, S. Plagenhoef (Prentice-Hall, Englewood Cliffs, NJ, 1971). A classic treatise summarizing the use and results of cinematographic techniques to analyze basic motions. (I)

2. **The Physics of Ball Games**, C. B. Daish (English U. P., London, 1972). Easily the most rigorous, clearly written book available concerning the physics of ball games, with examples from golf, cricket, and billiards. Written in two parts, the first contains a "general treatment of the subject, of interest to sportsmen in general," and the second "includes the full mathematical treatment for coaches and physicists in particular." (I)

3. **Mechanics and Sport, AMD Volume 4**, edited by J. L. Bleustein (American Society of Mechanical Engineers, New York, 1973). A collection of 20 research papers from an engineering symposium.

4. **A Bibliography of Biomechanics Literature**, J. G. Hay (published privately, 3056 Muscatine Ave., Iowa City, IA 52240, 1981). An exhaustive collection of several thousand references to journal and dissertation literature, organized into more than 100 separate bibliographies. Anyone who is thinking about performing an experiment or writing a research paper about the physics of sports should obtain this book. (A)

5. **The Dynamics of Sports—Why That's The Way the Ball Bounces**, D. F. Griffing (Mohican, Loudonville, OH, 1982). A textbook for an introductory physics course entitled "Physics in Sports," intended for nonscience majors. (E)

6. **Bicycling Science (2nd ed.)**, F. R. Whitt and D. G. Wilson (MIT, Cambridge, MA, 1982). A remarkable book exploring the past, present, and future approaches to the problem of human powered transportation, from the viewpoint of the engineer or physicist. (I)

7. **Sport Science—Physical Laws and Optimum Performance**, P. J. Brancazio (Simon and Schuster, New York, 1984). A readable, nontechnical approach using only elementary equations, this book explains appropriate concepts of basic physics to sports enthusiasts. (E)

7(a). **Mathematics in Sport**, M. S. Townend (Halsted, New York, 1984). An elementary physics analysis of a broad spectrum of activities and sports, originally written for a mathematics course for undergraduate physical education majors. The appendix includes several interactive BASIC programs for evaluating selected sports phenomena. (E)

8. **Newton at the Bat: The Science in Sports**, edited by E. W. Schrier and W. F. Allman (Scribner's, New York, 1984). A collection of essays of variable interest to physicists, all reprinted from **Science 80–Science 84**. (E)

9. **Physics of Human Motion**, J. W. Halley (Burgess, Minneapolis, MN, 1984). An introductory physics textbook designed for physical education and dance majors. (E)

10. **The Biomechanics of Sports Techniques (3rd ed.)**, J. G. Hay (Prentice-Hall, Englewood Cliffs, NJ, 1985). A widely used textbook for introductory graduate courses in biomechanics. It presents information about a wide variety of sports topics and provides a fairly extensive set of references. (E)

IV. ARTICLES ABOUT PARTICULAR SPORTS

A. Archery

11. **"Physics of bow and arrows,"** P. E. Klopsteg, Am. J. Phys. **11**, 175 (1943). (E)

12. **"Ballistics of the modern-working recurve bow and arrow,"** B. G. Schuster, Am. J. Phys. **37**, 364 (1969). (I)

13. **"An optimally designed archery,"** T. Soong, in *Mechanics and Sport, AMD-Vol. 4*, edited by J. L. Bleustein (American Society of Mechanical Engineers, New York, 1973), pp. 85–100. (A)

14. **"Bow and arrow dynamics,"** W. C. Marlow, Am. J. Phys. **49**, 320 (1981). (I)

B. Baseball

15. **"Effect of spin and speed on the lateral deflection (curve) of a baseball; and the Magnus effect for smooth spheres,"** L. J. Briggs, Am. J. Phys. **27**, 589 (1959). (I)

16. **"An analysis of the aerodynamics of pitched baseballs,"** C. Selin, Res. Q. **30**, 232 (1959). (I)

17. **"Batting the ball,"** P. Kirkpatrick, Am. J. Phys. **31**, 606 (1963). (I)

18. **"Home run hitting,"** J. D. Patterson, Phys. Teach. **5**, 167 (1967). (E)

19. **"Catching a baseball,"** S. Chapman, Am. J. Phys. **36**, 868 (1968). (I)

20. **"Aerodynamics of a knuckleball,"** R. G. Watts and E. Sawyer, Am. J.

Phys. **43**, 960 (1975). (I)

21. **"Aerodynamic drag crisis and its possible effect on the flight of baseballs,"** C. Frohlich, Am. J. Phys. **52**, 325 (1984). (I)

22. **"The sweet spot of a baseball bat,"** H. Brody, Am. J. Phys. **54** (to be published, 1986). (I)

23. **"Trajectory of a fly ball,"** P. J. Brancazio, Phys. Teach. **23**, 20 (1985). (I)

24. **"Looking into Chapman's homer: The physics of judging a fly ball,"** P. J. Brancazio, Am. J. Phys. **53**, 849 (1985). (I)

C. Basketball

*25. **"Physics of basketball,"** P. J. Brancazio, Am. J. Phys. **49**, 356 (1981); also, Am. J. Phys. **50**, 567, 944 (1982). (I)

26. **"Kinematics of the free throw in basketball,"** A. Tan and G. Miller, Am. J. Phys. **49**, 542 (1981). (I)

D. Bowling

27. **"Bowling frames: Paths of a bowling ball,"** D. C. Hopkins and J. D. Patterson, Am. J. Phys. **45**, 263 (1977). (I)

*28. **"On the dynamics of a weighted bowling ball,"** R. L. Huston, C. Passerello, J. M. Winget, and J. Sears, J. Appl. Mech. **46**, 937 (1979). (A)

E. Cycling

29. **"The stability of the motion of a bicycle,"** F. J. W. Whipple, Q. J. Pure Appl. Math. **30**, 312 (1899). (I)

30. **"The stability of the bicycle,"** D. E. H. Jones, Phys. Today **23** (4), 34 (1970). (E)

31. **"Dynamics of a bicycle: Nongyroscopic aspects,"** J. Liesegang and A. R. Lee, Am. J. Phys. **46**, 130 (1978). (I)

*32. **"Reduction of wind resistance and power output of racing cyclists and runners traveling in groups,"** C. R. Kyle, Ergonomics **22**, 387 (1979). (I)

33. **"Some nonexplanations of bicycle stability,"** D. Kirshner, Am. J. Phys. **48**, 36 (1980). (E)

34. **"The stability of bicycles,"** J. Lowell and H. D. McKell, Am. J. Phys. **50**, 1106 (1982). (I)

35. **"The aerodynamics of human-powered land vehicles,"** A. C. Gross, C. R. Kyle, and D. J. Malewicki, Sci. Am. **249** (6), 142 (1982). (E)

36. **"Improving the racing bicycle,"** C. R. Kyle and E. Burke, Mech. Eng. **106** (9), 34 (1984). (I)

F. Golf

37. **"On the path of a rotating spherical projectile,"** P. G. Tait, Trans. R. Soc. Edinburgh **37**, 427 (1893). (I)

*38. **"The dynamics of a golf ball,"** J. J. Thomson, Nature **85**, 2147 (1910). (I)

39. **"The aerodynamics of golf balls,"** J. M. Davies, J. Appl. Phys. **20**, 821 (1949). (I)

40. **"On the dynamics of the swing of a golf club,"** T. Jorgensen, Am. J. Phys. **38**, 644 (1970). (I)

41. **"Golf ball aerodynamics,"** P. W. Bearman and J. K. Harvey, Aeronaut. Q. **27**, 112 (1976). (I)

42. **"Maximum projectile range with drag and lift, with particular application to golf,"** H. Erlichson, Am. J. Phys. **51**, 357 (1983). (I)

G. Gymnastics

43. **"Photographs of a tumbling cat,"** Editor, Nature **51**, 80 (1894). (E)

44. **"A dynamical explanation of the falling cat phenomenon,"** T. R. Kane and M. P. Scher, J. Solids Struct. **5**, 663 (1969). (A)

45. **"Human self-rotation by means of limb movements,"** T. R. Kane and M. P. Scher, J. Biomech. **3**, 39 (1970). (A)

46. **"A computational technique to determine the angular momentum of a human body,"** J. G. Hay, B. D. Wilson, J. Dapena, and G. G. Woodworth, J. Biomech. **10**, 269 (1977). (I)

*47. **"Do springboard divers violate angular momentum conservation?,"** C. Frohlich, Am. J. Phys. **47**, 583 (1979). (I)

48. "The physics of somersaulting and twisting," C. Frohlich, Sci. Am. **242**, 154 (1980). (E)

49. "An analysis of the rotational stability of the layout back somersault," P. A. Lightsey, Am. J. Phys. **51**, 115 (1983). (I)

H. Martial arts

50. "Karate strikes," J. D. Walker, Am. J. Phys. **43**, 845 (1975). (I)

51. "Physics and the art of kicking and punching," H. Blum, Am. J. Phys. **45**, 61 (1977). (I)

52. "The physics of karate," M. S. Feld, R. E. McNair, and S. R. Wilk, Sci. Am. **240**, 150 (1979). (E)

53. "The physics of karate," S. R. Wilk, R. E. McNair, and M. S. Feld, Am. J. Phys. **51**, 783 (1983). (I)

I. Racket sports

54. "The irregular flight of a tennis ball," Lord Rayleigh, Messenger Math. **7**, 14 (1877). (I)

*55. "Physics of the tennis racket," H. Brody, Am. J. Phys. **47**, 482 (1979). (I)

56. "Terminal velocity of a shuttlecock in vertical fall," M. Peastrel, R. Lynch, and A. Armenti, Am. J. Phys. **48**, 511 (1980). (I)

57. "Exercise in probability and statistics, or the probability of winning at tennis," G. Fischer, Am. J. Phys. **48**, 14 (1980). (I)

*58. "Physics of the tennis racket II: The 'sweet spot'," H. Brody, Am. J. Phys. **49**, 816 (1981). (I)

59. "Mathematical modeling and simulation of a tennis racket," M. Brannigan and S. Adlai, Med. Sci. Sports Exercise **13**, 44 (1981). (A)

60. "Tennis: the influence of grip tightness on reaction time and rebound velocity," B. C. Elliot, Med. Sci. Sports Exercise **14**, 348 (1982). (I)

61. "Mechanical analysis of racket and ball during impact," Y. K. Liu, Med. Sci. Sports Exercise **15**, 388 (1983). (A)

62. "Resultant tennis ball velocity as a function of off-center impact and grip firmness," M. D. Grabiner, J. L. Groppel, and K. R. Campbell, Med. Sci. Sports Exercise **15**, 542 (1983). (I)

*63. "A mechanical analysis of a special class of rebound phenomena," J. L. Andrews, Med. Sci. Sports Exercise **15**, 256 (1983). (A)

64. "That's how the ball bounces," H. Brody, Phys. Teach. **22**, 494 (1985). (I)

J. Track and field

65. "Behavior of the discus in flight," J. A. Taylor, Athletic J. **12**(4), 9, 45 (1932). (E)

66. "Bad physics in athletic measurement," P. Kirkpatrick, Am. J. Phys. **12**, 7 (1944). (E)

67. "Aerodynamic and mechanical forces in discus flight," R. V. Ganslen, Athletic J. **68**(4), 50, 68, 88 (1964). (I)

68. "The influence of wind resistance in running and walking and the mechanical efficiency of work against horizontal or vertical forces," L. G. C. E. Pugh, J. Physiol. **213**, 255 (1971). (I)

69. "A theory of competitive running," J. B. Keller, Phys. Today **26**, 43 (1973). (I)

70. "The dynamics of the javelin throw," T. C. Soong, J. Appl. Mech. **42**, 257 (1975). (A)

71. "The dynamics of the discus throw," T. C. Soong, J. Appl. Mech. **43**, 531 (1976). (A)

72. "Maximizing the range of the shot put," D. B. Lichtenberg and J. G. Wills, Am. J. Phys. **46**, 546 (1978). (I)

73. "Newtonian mechanics and the human body: Some estimates of performance," H. Lin, Am. J. Phys. **46**, 15 (1978). (E)

74. "Fast running tracks," J. A. McMahon and P. R. Greene, Sci. Am. **239**, 148 (1978). (E)

*75. "The influence of track compliance on running," T. A. McMahon and P. R. Greene, J. Biomech. **12**, 893 (1979). (I)

76. "Dynamics of the pole vault," M. Hubbard, J. Biomech. **13**, 965 (1980). (A)

77. "Physics of sprinting," I. Alexandrov and P. Lucht, Am. J. Phys. **49**, 254 (1981). (I)

*78. "Aerodynamic effects on discus flight," C. Frohlich, Am. J. Phys. **49**, 1125 (1981). (I)

79. "The influence of aerodynamic and biomechanical factors on long jump performance," A. J. Ward-Smith, J. Biomech. **16**, 655 (1983). (I)

80. "A parametric solution to the elastic pole-vaulting pole problem," G. M. Griner, J. Appl. Mech. **51**, 409 (1984). (A)

81. "On 'waddling' and race walking," O. Helene, Am. J. Phys. **52**, 656 (1984). (I)

82. "Simulation of javelin flight using experimental aerodynamic data," M. Hubbard and H. J. Rust, J. Biomech. **17**, 769 (1984). (I)

83. "Optimal javelin trajectories," M. Hubbard, J. Biomech. **17**, 777 (1984). (I)

84. "Javelin dynamics with measured lift, drag, and pitching moments," M. Hubbard and H. J. Rust, J. Appl. Mech. **51**, 406 (1984).

85. "Air resistance and its influence on the biomechanics and energetics of sprinting at sea level and at altitude," A. J. Ward-Smith, J. Biomech. **17**, 339 (1984). (I)

86. "Effect of wind and altitude on record performance in foot races, pole vault, and long jump," C. Frohlich, Am. J. Phys. **53**, 726 (1985). (I)

*87. "A mathematical theory of running, based on the first law of thermodynamics, and its application to the performance of world-class athletes," A. J. Ward-Smith, J. Biomech. **18**, 337 (1985). (I)

88. "A mathematical analysis of the influence of adverse and favourable winds on sprinting," A. J. Ward-Smith, J. Biomech. **18**, 351 (1985). (I)

K. Winter sports

89. "The physics of ski turns," J. I. Shonle and D. L. Nordich, Phys. Tech. **10**, 491 (1972). (I)

90. "Control systems approach to a ski-turn analysis," J. M. Morawski, J. Biomech. **6**, 267 (1973). (A)

91. "Biomechanics of optimal flight in ski-jumping," L. P. Remizov, J. Biomech. **17**, 161 (1984). (I)

92. "The influence of air friction on speed skating," G. J. van Ingen Shenau, J. Biomech. **15**, 449 (1982). (I)

92(a). "On the motion of an ice hockey puck," K. Voyenli and E. Eriksen, Am. J. Phys. **53**, 1149 (1985). (I)

L. Other

93. "Some hydrodynamic aspects of rowing," J. F. Wellicome, in *Rowing—A Scientific Approach*, edited by J. G. P. Williams and A. C. Scott (A. S. Burns, New York, 1967), pp. 22–63. (I)

94. "Kinematics of an ultraelastic rough ball," R. L. Garwin, Am. J. Phys. **37**, 88 (1969). (I)

95. "On the dynamics of men and boats and oars," D. L. Pope, in *Mechanics and Sport, AMD–Vol. 4*, edited by J. L. Bleustein (American Society of Mechanical Engineers, New York, 1973), pp. 113–130. (A)

96. "Computerized biomechanical analysis of human performance," G. Ariel, in *Mechanics and Sport, AMD–Vol. 4*, edited by J. L. Bleustein (American Society of Mechanical Engineers, New York, 1973), pp. 267–275. (I)

97. "The swing of a cricket ball," J. H. Horlock, in *Mechanics and Sport, AMD–Vol. 4*, edited by J. L. Bleustein (American Society of Mechanical Engineers, New York, 1973), pp. 293–303. (I)

98. "Lateral dynamics and stability of the skateboard," M. Hubbard, J. Appl. Mech. **46**, 931 (1979). (I)

99. "Aerodynamics of the cricket ball," R. Mehta and D. Wood, New Sci. **87**, 442 (1980). (E)

100. "The physics of kicking a football," P. J. Brancazio, Phys. Teach. **23**, 403 (1985). (I)

Reprinted from *American Journal of Physics*, **27**. ©1959 American Association of Physics Teachers.

Effect of Spin and Speed on the Lateral Deflection (Curve) of a Baseball; and the Magnus Effect for Smooth Spheres

Lyman J. Briggs

National Bureau of Standards, Washington, D. C.

(Received March 26, 1959)

The effect of spin and speed on the lateral deflection (curve) of a baseball has been measured by dropping the ball while spinning about a vertical axis through the horizontal wind stream of a 6-ft tunnel. For speeds up to 150 ft/sec and spins up to 1800 rpm, the lateral deflection was found to be proportional to the spin and to the square of the wind speed. When applied to a pitched ball in play, the *maximum* expected curvature ranges from 10 to 17 in., depending on the spin. The deflections of rough baseballs accord in direction with that predicted by the Magnus effect. But with *smooth* balls at low speeds the deflection is in the *opposite* direction. This is studied with an apparatus specially designed to measure the pressure at any point in the equatorial plane of the rotating ball.

INTRODUCTION

EVERYONE who has played golf or baseball or tennis knows that when a ball is thrown or struck so as to make it spin, it usually "curves" or moves laterally out of its initial vertical plane.

How is this lateral deflection related to the spin and speed of the ball? An experimental answer to this question was sought for baseballs.

The first explanation of the lateral deflection of a spinning ball is credited by Lord Rayleigh[1] to Magnus,[2] from whom the phenomenon derives its name, the "Magnus effect."

The commonly accepted explanation is that a spinning object creates a sort of whirlpool of rotating air about itself. On the side where the motion of the whirlpool is in the same direction as that of the wind stream to which the object is exposed, the velocity will be enhanced. On the opposite side, where the motions are opposed, the velocity will be decreased. According to Bernoulli's principle, the pressure is lower on the side where the velocity is greater, and consequently there is an unbalanced force at right angles to the wind. This is the Magnus force.

In the case of a cylinder or a sphere, the so-called whirlpool, or more accurately the circulation, does not consist of air set into rotation by friction with the spinning object. Actually an object such as a cylinder or a sphere can impart a spinning motion to only a very small amount of air, namely to that in a thin layer next to the surface. It turns out, however, that the motion imparted to this layer affects the manner in which the flow separates from the surface in the rear, and this in turn affects the general flow field about the body and consequently the pressure in accordance with the Bernoulli relationship. The Magnus effect arises when the flow follows farther around the curved surface on the side traveling with the wind than on the side traveling against the wind. This phenomenon is influenced by the conditions in the thin layer next to the body, known as the boundary layer, and there may arise certain anomalies in the force if the spin of the body introduces anomalies in the layer, such as making the flow turbulent on one side and not on the other. As we shall see, a reverse Magnus effect may occur for smooth spheres. Rough balls, such as baseballs and tennis balls, do not show this anomalous effect.

The ingenious experiments which led Magnus to the discovery of the effect were made chiefly with a small cylinder rotating about a vertical axis in a horizontal wind and so mounted that it was free to move laterally across the wind, but not downstream. The pull of a cord wrapped around the axis served to give the cylinder its initial spin. Magnus makes no comment about the smoothness of the surface of the cylinder. The boundary-layer concept, which was introduced by Prandtl in 1904, was of course not available to Magnus.

[1] Lord Rayleigh, "*On the irregular flight of a tennis ball*," Scientific Papers 1, 344 (1869–81).

[2] G. Magnus, "*On the deviation of projectiles; and on a remarkable phenomenon of rotating bodies.*" Memoirs of the Royal Academy, Berlin (1852). English translation in Scientific Memoirs, London (1853), p. 210. Edited by John Tyndall and William Francis.

590   L Y M A N   J .   B R I G G S

The beautiful photograph made by Professor F. N. M. Brown of the University of Notre Dame in his low-turbulence wind tunnel illustrates what has just been said (see Fig. 1). Here the wind with its smoke filaments is coming from the right at 60 feet per second. The ball is stationary in the tunnel but spinning counterclockwise at 1000 revolutions per minute about a horizontal axis at right angles to the wind. The crowding together of the smoke filaments over the *top* of the ball shows an increased velocity in this region and a corresponding decrease in pressure, which according to the Bernoulli principle, would tend to deflect the ball *upward* across the wind stream.

It will also be noted that the *wake* of the ball has been deflected *downward*. According to the principle of the conservation of momentum, this must likewise be accompanied by a corresponding *upward* thrust on the ball.

Put in other words, if the wind speed is from east to west and the ball is spinning counterclockwise about a *vertical* axis, then the Magnus force on the ball is directed towards the north.

For a further discussion of the flow past rotating cylinders, including many photographs, see Prandtl[3] and Goldstein.[4]

PART I. EXPERIMENTS WITH BASEBALLS

Air-Gun Experiments

My first measurements were made with an air-gun which had earlier been constructed at the National Bureau of Standards to measure the coefficient of restitution of baseballs.[5] The ball was mounted on a spinning tee located in front of the muzzle. The wooden projectile from the air-gun drove the spinning ball a distance of 60 ft (the distance from the pitcher's rubber to the home plate) where it made an imprint on a vertical target.

The spin of the ball before impact was measured with a Strobotac. The speed could (in theory) be computed by measuring the drop of the ball, i.e., the vertical distance of the projected

horizontal axis of the gun above the point of impact.

The direction of the spin about the vertical axis could be reversed at will and measurements were made with the ball first spinning clockwise (looking down) and then counterclockwise. One-half of the horizontal distance between the two target imprints gave the lateral displacement sought.

This setup was simple, and at first sight appeared usable. The observed deflections were in the expected direction, and shifted to the other side of the target when the spin was reversed. But the results were erratic. A stroboscopic camera was then installed 30 ft above the floor looking down on the last half of the flight path, and the position of the ball was photographed at exact 0.05-sec intervals against the scale on the floor.

These photographic measurements gave the trajectory of the ball, its speed and the drag; but they also indicated that the spin of the ball was greatly reduced when it was distorted by the impact of the projectile; and that the reaction between the spinning ball and the projectile gave rise to a small component of velocity normal to the flight path, which contributed to the observed lateral deflections. The trajectory was in fact that of a batted ball instead of a pitched ball. This line of attack was consequently abandoned in favor of wind tunnel measurements.

Wind Tunnel Measurements

In these measurements, the spinning ball was dropped from the upper side of the NBS 6-ft

FIG. 1. Showing airflow past spinning ball in wind tunnel. Wind coming from right, 60 ft/sec. Spin 1000 rpm, counterclockwise, about a horizontal axis at right angles to wind. Magnus force, upward. Courtesy of Professor F. N. M. Brown, University of Notre Dame.

[3] Ludwig Prandtl, *Essentials of Fluid Dynamics* (Hafner Publishing Company, Inc., New York, 1952).
[4] S. Goldstein, *Modern Developments in Fluid Dynamics* (Clarendon Press, Oxford, England, 1938), Vols. 1 and 2.
[5] Lyman J. Briggs, J. Research Natl. Bur. Standards **34**, 1 (1945).

LATERAL DEFLECTION OF A BASEBALL 591

octagonal wind tunnel across a horizontal wind of known velocity. By coating the bottom of the ball lightly with a lubricant containing lamp-black, its point of impact was recorded on a sheet of cardboard fastened to the tunnel floor. The lateral deflection, which is of immediate interest, was taken as one-half of the measured spread of the two points of impact, with the ball spinning first clockwise and then reversed.

The spinning mechanism was mounted outside on the top of the tunnel with its hollow shaft projecting vertically downward one-half inch through the tunnel wall. A concentric suction cup to support the ball was mounted on this shaft. The spinning ball was released by a quick-acting valve which cut off the suction and opened the line to the atmosphere.

The spinner was belt-driven by a small dc motor with its armature current supplied from a potentiometer circuit to secure the desired speed range. The angular speed (rpm) was measured with a calibrated Strobotac which illuminated a rotating target on the spinner mechanism. The ball was shielded by a thin-walled cylinder (4 in. o.d., 4 in. long) mounted on the inner wall of the tunnel, concentric with the spinner shaft. While this introduced some additional turbulence over that created by the bare ball on its spinner, it gave more consistent

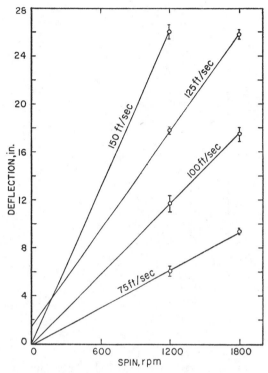

FIG. 2. Lateral deflection of a baseball, spinning about a vertical axis, when dropped across a horizontal windstream. These values are all for the same time interval, 0.6 sec, the time required for the ball to cross the stream.

results, and prevented irregularities caused by the ball being torn off its support by the wind stream during release. Official American League balls were used throughout the measurements.

Owing to the method of construction, the center of gravity of a baseball often does not coincide exactly with its geometrical center. As a result of this asymmetry, the ball rotating on its spinner is subject to a lateral centrifugal force. This causes the ball to depart from a truly vertical fall when there is no wind. The departure may be upstream, laterally, or downstream, depending on the angular position of the heavy side of the ball when released, and results in a scatter of impacts for the same spin and windspeed.

To minimize this effect, the ball was turned in different positions while being placed in the suction cup of the spinner, until a position in which the center of gravity appeared to fall on the spin axis was found by trial.

At least three measurements were made for each spin and wind speed, as well as for the spin reversed. The mean values are given in Table I

TABLE I. Lateral deflection of a spinning baseball in a 6-ft drop across the tunnel windstream at various spins and speeds.

Spin rpm	Speed ft/sec	Deflection, inches	Ratio of deflections	Ratio of (speeds)²
1200	125	17.8	1.52	1.56
	100	11.7		
1200	150	26.0	1.46	1.44
	125	17.8		
1200	150	26.0	2.22	2.25
	100	11.7		
1200	100	11.7	1.92	1.77
	75	6.1		
1200	125	17.8	2.91	2.79
	75	6.1		
1200	150	26.0	4.25	4.0
	75	6.1		
1800	125	25.8	1.47	1.56
	100	17.5		
1800	125	25.8	2.98	2.79
	75	9.4		
1800	100	175	1.81	1.77
	75	94		

592   L Y M A N   J . B R I G G S

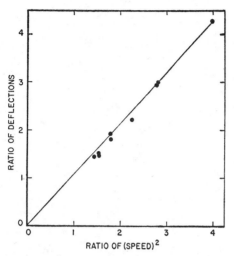

FIG. 3. Graph showing that the observed lateral deflections are proportional to the *square* of the wind speed.

and shown graphically in Fig. 2 together with the standard deviations. The lateral deflection in all instances accorded in direction with that expected from the Magnus effect.

It will be noted from Fig. 2 that straight lines drawn through the observed deflections at spins of 1200 and 1800 rpm for different wind speeds pass nearly through the origin. In other words, within experimental limits, the lateral deflection is directly proportional to the spin.

The effect of wind speed on the lateral deflection is shown in Table I. The fourth column of the table gives the ratio of the observed deflections at known wind speeds. For comparison, the last column gives the ratio of the corresponding wind speeds squared. These results are plotted in Fig. 3. Subject to experimental errors, the square relationship is seen to hold.

We conclude then that for speeds up to 150 ft/sec and spins up to 1800 rpm, the lateral deflection of a baseball spinning about a vertical axis is directly proportional to spin and to the square of the wind speed.

Maximum Curve Expected for Pitched Baseballs in Play

All the lateral deflections shown in Fig. 2 refer to what took place in 0.6 sec, the time required for the ball to fall across the wind stream of the 6-ft tunnel. We have now to convert these measurements into what deflection would be expected if the ball were traveling the 60 ft from the pitcher's rubber to the plate at various speeds and spins.

The results are summarized in Table II. A ball thrown at a speed of 100 ft/sec would travel the 60 ft from rubber to plate in 0.6 sec, which is the time required for the ball to fall across the wind stream in the tunnel measurements. Consequently in this case the observed lateral deflection in the tunnel would be equal to that of the ball in play. In other cases a correction is necessary.

The lateral deflecting force of the tunnel stream is practically constant for a given spin and speed, so that the lateral acceleration of the ball is constant and the distance traveled is proportional to the square of the elapsed time. At 125 ft/sec, the thrown ball travels 60 ft in 0.48 sec, while the dropped ball takes 0.6 sec to cross the tunnel. Hence the deflection in 0.48 sec would be

$$17.8 \text{ in.} \times (0.48/0.60)^2 = 11.4 \text{ in.}$$

It will be noted from Table II that the amount the ball curves in 60 ft is proportional to the spin, but is practically independent of the speed, namely, about 11 in. at 1200 rpm and 17 in. at 1800 rpm. This result may seem surprising until it is recalled that at the higher speeds, the ball is in the 60-ft zone for a shorter time and that the lateral displacement is proportional to the time squared.

These measurements were all made with the ball spinning about a vertical axis, which gave the maximum lateral deflection. Usually in play the spin axis is inclined, which reduces the effect. If the spin axis were horizontal and normal to the flight path, no lateral deflection would take

TABLE II. Curving of baseballs in play.

Speed ft/sec	Spin rev/sec	Turns in 60 ft	Curve (lat. def.) in., in 60 ft
75	20	16	10.8
75	30	24	16.7
100	20	12	11.7
100	30	18	17.5
125	20	9.6	11.4
125	30	14.4	16.5
150	20	8	11.6

place. With clockwise spin (seen from right) the pitch would be a drop.

Notes Pertaining to Baseballs in Play

These measurements were designed to cover simply the range of conditions encountered in play. The following records are of interest in this connection. Bob Feller, former Cleveland pitcher, in 1947 threw a baseball across the plate at a speed of 98.6 mph (144 ft/sec), as measured with electronic instruments. J. G. Taylor Spink, Editor of the *Sporting News*, states that this is the accepted world record for the fastest pitch.[6] The next fastest pitch of record is 94.7 mph (138 ft/sec) by Atley Donald, former New York Yankee pitcher, in 1939.

Dr. H. L. Dryden kindly arranged for the measurement of the "terminal velocity" of a baseball, that is its maximum speed after falling from a great height to the ground.[7] This measurement was carried out in a vertical wind-tunnel at the National Advisory Committee for Aeronautics, the wind speed being adjusted until the ball just floated in the windstream. The terminal velocity was about 140 ft/sec. The celebrated catch by Charles Street, of the Washington Ball Club, of a ball dropped from a window of the Washington Monument gave a computed velocity *in vacuo* of 179 ft/sec. The NACA measurements show that, owing to the resistance of the air, the actual speed could not have exceeded 140 ft/sec. However, home runs *batted* into the stands must have an *initial* velocity considerably higher than this.

With the cooperation of the pitchers of the Washington Ball Club, the spin of a pitched ball was measured. This was done by fastening one end of a long tape to the ball and then laying the rest loosely (but untwisted) on the ground between the rubber and the plate, the free end being pegged down. After the ball was caught, the number of turns was counted. The highest spins measured were 15.5–16 turns in 60 ft and the lowest 7–8. Assuming the speed of the pitch to be 100 ft/sec, the maximum spin measured would be about 1600 rpm. These spins are covered by the wind tunnel observations.

PART 2. SMOOTH BALLS AND THE MAGNUS EFFECT

The experiments with *rough* baseballs (Part 1) all showed lateral deflections in conformance with the Magnus effect. But with *smooth* balls, especially at low wind speeds, an anomaly is encountered. The deflections are usually in the *opposite* direction.

Maccoll, using a 6-in. diam wooden sphere, rotating on a wind tunnel balance, was apparently the first to demonstrate the existence of a small negative lift on a sphere at low wind speeds.[8] His C_L curve, negative at first, crosses the *no lift* axis when the equatorial speed/wind speed $= U/V = 12.3/24.6 = 0.5$.

Davies, in his experiments with smooth and dimpled golf balls, found that for the smooth ball the lift was negative at all rotational speeds below 5000 rpm (equatorial speed, 2100 ft/min).[9] Above this, the lift was positive, but was less than for the standard ball.

In Davies' measurements, the ball was dropped across the horizontal wind stream of an open tunnel operating at 105 ft/sec. The axis of spin was horizontal and normal to the wind stream. A quick-acting device released the spinning ball. The vertical drop was 0.67 to 1.3 ft. Spins up to 8000 rpm could be obtained.

It will be noted that in Davies' experiments the point of impact on the tunnel floor represents the combined effect of spin and drag, both being directed downstream. To get the lift, the point of impact had to be reduced by the drag with no spin; whereas in my baseball measurements the lateral deflection, being at right angles to the drag, could be measured directly.

Brown has recently measured the lift coefficient of a sphere of 3.36-in. diam mounted on a balance in a low turbulence tunnel, and rotating at fixed speeds ranging from 700 to 4500 rpm.[10] His results all show a negative lift at low wind speeds, the graphs crossing the no-lift axis at values of V/U ranging from 0.1 to 0.5, where U is the wind speed and V is the peripheral speed.

---

[6] J. G. Taylor Spink (personal communication).

[7] H. L. Dryden (personal communication, with permission).

[8] Maccoll, J. Roy. Aeronaut. Soc. **32**, 777 (1928).

[9] J. M. Davies, J. Appl. Phys. **20**, 821 (1949). This paper contains references to articles not here recorded.

[10] F. N. M. Brown, University of Notre Dame, Notre Dame, Indiana (personal communication, 1958; unpublished data, with permission).

594   L Y M A N   J .   B R I G G S

TABLE III. Lateral deflection (inches) of smooth Bakelite ball, 6-ft drop.

Wind speed, ft/sec	Spin, rpm	Deflection, in.	
75	1200	1.1	Pro-Magnus
100	1200	0.5	Pro-Magnus
125	1200	0.6	Anti-Magnus
150	1200	2.2	Anti-Magnus
75	1800	1.4	Pro-Magnus
100	1800	0.5	Pro-Magnus
125	1800	0.8	Anti-Magnus
150	1800	7.3	Anti-Magnus

Smooth Rubber Ball

The writer has measured the lateral deflection of a *smooth* rubber ball, using the same setup employed with baseballs. The ball, which had been cast in an accurately spherical mold, was 2.88 in. in diam, practically that of a baseball, but was heavier (wt 188 g; baseball about 145 g). Its center of gravity agreed closely with its geometrical center and it had a good bounce.

This smooth ball was deflected laterally *opposite* in direction to that of the baseballs. The deflection was small (partly owing to the increased weight) but increased steadily from 3.6 in. at 75 ft/sec (1800 rpm) to 8.8 in. at 150 ft/sec (all negative Magnus). At 1800 rpm the rotational velocity of a point on the equator was 1350 ft/min. Davies states that his smooth golf ball began to develop a positive lift at 5000 rpm (equatorial speed, 2100 ft/min).

Bakelite Sphere

Similar tests were made with a smooth Bakelite sphere (good ground finish) of 3-in. diam, sphericity 0.005 in., wt 312 g. The Reynolds number at 150 ft/sec was 2.4×10^5. At this speed the ball was presumably passing out of the critical Reynolds-number range for the drag coefficient for spheres.[11] See Goldstein.[4]

The lateral deflections, which are very small, are given in Table III. The unexpected result was that at the lower wind speeds the ball had a Magnus deflection, crossing to the anti-Magnus regime between 100 and 125 ft/sec. Here we seem to have a new effect for *smooth* balls, which has been masked in the measurements reported above.

It seems now to be well established by the results of four different laboratories that a spinning *smooth* ball at low wind-tunnel speeds usually does not conform with Magnus effect but exerts a "negative" lift. It appears likely that this occurs in a certain Reynolds-number range where it is possible for the boundary layer on the side moving with the wind to remain laminar while that on the opposite side becomes turbulent. Since a turbulent layer will in general follow farther around the surface before separating than a laminar layer, the crosswind force is reversed, if the rate of spin is not too high. We would expect to pass from this condition into the Magnus regime when the spin is sufficiently increased or when the flow on both sides becomes turbulent, as it will when the Reynolds number is sufficiently increased.

Evidently we may also pass into the Magnus regime at low Reynolds numbers by arranging conditions so that the flow is laminar on both the approaching and receding sides, the requirements now being that disturbances, such as vibration of the sphere and turbulence in the wind stream, are sufficiently reduced. This is believed to be the explanation of the pro-Magnus results of Table III.

In general, then, we find that the sign and magnitude of the effect of spin are dependent on dynamic conditions as well as on the "smooth-

---

[11] The Reynolds number is the product of the speed of the wind and the diameter of the sphere divided by the kinematic viscosity of the air.

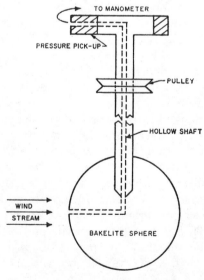

FIG. 4. Schematic drawing of apparatus used to measure the pressure at various points on a ball spinning in a wind stream.

ness" of the ball. The foregoing explanations have been offered as the most reasonable ones on the basis of the information available. The author knows of no detailed investigations of the flow in these cases.

PART 3. PRESSURE DISTRIBUTION OVER A SPINNING SPHERE IN ITS EQUATORIAL PLANE

To learn more about the forces acting on the surface of a smooth rotating sphere, the following apparatus, shown schematically in Fig. 4, was constructed.

The smooth Bakelite sphere (3-in. diam) used earlier in the free-fall measurements was mounted on a hollow vertical shaft ($\frac{3}{8}$-in. diam), the center of the unshielded sphere projecting downward ($2\frac{1}{2}$ in.) into the horizontal wind stream of the tunnel. A $\frac{1}{8}$-in. hole drilled along an equatorial radius connected the surface of the sphere with the hollow shaft.

The shaft extended upward through the top wall of the 6-ft octagonal tunnel, where it was expanded into a cylindrical head ($1\frac{1}{8}$-in. diam), which was drilled radially to lie directly above the hole in the sphere. The head was surrounded by a closely fitted pickup sleeve, which through a drilled radial hole ($\frac{1}{16}$-in. diam) connected the ball with the manometer. The pickup sleeve could be rotated and held independently in any angular position about its vertical axis, so that the pressure on the ball at any point in its horizontal equatorial plane could be measured.

The oppoiste leg of the manometer was connected to a static pressure orifice in the wall of the tunnel. The wind speeds (75 and 125 ft/sec) were based on impact pressure measurements in the empty tunnel. The hollow shaft was belt-driven by a dc motor on a potentiometer circuit. The spin, measured by a stroboscope, was roughly 1800 rpm.

As far as the writer knows, this is the first time the pressure at the equatorial surface of a spinning sphere has been measured directly. It provides a point-to-point determination of the pressure around the entire periphery of the sphere and eliminates the disturbance set up by the introduction of an external probe. By drilling the entrance hole in the sphere at higher latitudes, the pressure distribution over the entire

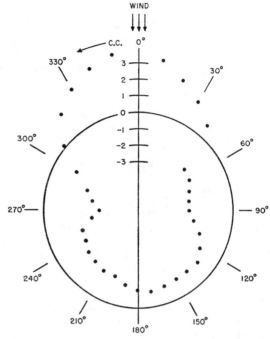

FIG. 5. Polar diagram of pressure distribution around smooth Bakelite ball (looking down along vertical spin axis). Ball, 3-in. diam; counter-clockwise spin, about 1800 rpm; wind speed, 125 ft/sec; pressure, inches of water.

surface of the sphere could be found. Only the equatorial pressures have been measured in the present study.

Two examples are given of the pressure distribution around the spinning sphere. In Fig. 5, the observed pressures at 125 ft/sec and 1800 rpm are shown in a polar diagram. It will be seen that the pressure is above atmospheric for 40°–50° on either side of the impinging windstream and that for this region the difference in pressure of symmetrical points would tend to force the ball to the right in Fig. 5, an anti-Magnus effect. The cosine projection of these above-atmospheric pressures on a horizontal diameter normal to the windstream is, however, small. For the remainder of the diagram, the pressures in the left half are consistently lower than corresponding points in the right half, tending to force the ball to the left, thus giving a positive Magnus effect.

If we sum up the pressure differences of corresponding points in Fig. 5 (after a cosine projection on a transverse equatorial diameter as illustrated in Fig. 6), we come out with a value of −5.78 pressure units, the minus sign indicating that the resultant transverse force is in the

596   L Y M A N   J .   B R I G G S

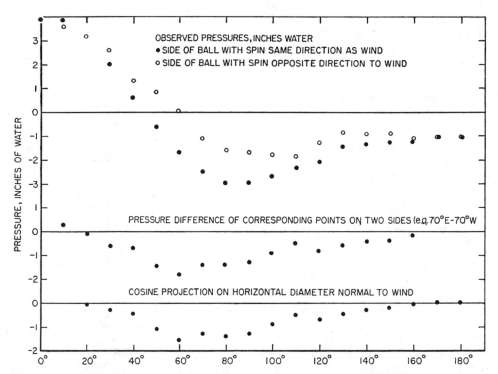

FIG. 6. Rough Bakelite ball, 3-in. diam; wind speed, 125 ft/sec; spin, about 1800 rpm. Meridional rubber bands stretched over ball to simulate roughness.

Magnus direction at 125 ft/sec and 1800 rpm. The corresponding measurement at 75 ft/sec is +1.77 pressure units, an anti-Magnus effect. These results are opposite in sign from those obtained when the Bakelite ball was dropped across the windstream. Here, however, we are dealing only with the pressures in a narrow equatorial belt on the ball ($\frac{1}{8}$-in. wide) and not with its entire surface.

Rough Bakelite Ball

Finally, the surface of the same ball used in the preceding measurements was roughened by attaching rubber bands along meridional lines. Figure 6 shows: (a) the observed pressures at 10° intervals measured from the direction of the wind; (b) the pressure difference of correspond-

ing points; and finally (c) the projection of these differences on a horizontal diameter normal to the wind. The resultant pressures are consistently in accord with the Magnus effect, but it will be noted that this effect is substantially reduced by what takes place on the opposite side of the ball. Their sum, −10.8, taken in conjunction with appropriate units of area, gives a measure of the resultant force on the ball at the equator. The corresponding figure for the smooth Bakelite ball at the same speed is −5.78.

ACKNOWLEDGMENT

I would like to express my indebtedness to G. B. Schubauer, B. L. Wilson, R. H. Heald, R. J. Hall, and G. H. Adams for valued assistance in various ways.

Physics of basketball

Peter J. Brancazio

Department of Physics, Brooklyn College, City University of New York, Brooklyn, New York 11210
(Received 29 November 1979; accepted 11 August 1980)

Does a knowledge of physics help to improve one's basketball skills? Several applications of physical principles to the game of basketball are examined. The kinematics of a basketball shot is studied, and criteria are established for determining the best shooting angle at any given distance from the basket. It is found that there is an optimum shooting angle which requires the smallest launching force and provides the greatest margin for error. Some simple classroom illustrations of Newtonian mechanics based on basketball are also suggested.

Reprinted from *American Journal of Physics*, **49**, 4. ©1981 American Association of Physics Teachers.

I. INTRODUCTION

Anyone who has followed sports closely over the past few decades cannot help but notice the evergrowing influence of science and technology. Athletes have benefited greatly from advances in sports medicine, training techniques, and equipment design. In a number of sports, detailed technical and statistical analyses of patterns of play have led to more sophisticated game strategies. What has yet to be fully developed, however, is an understanding and appreciation of the physical principles underlying various sports.

The scientific literature of the physics of sports is curiously sparse. In the past this Journal has contained articles on baseball,[1-4] tennis,[5] bowling,[6] golf,[7] springboard diving,[8] and (miraculously) only one brief article on jogging.[9] However, there seems to be no published research on the physics of basketball. Even within the field of physical education, remarkably little has been done on this subject; the *Research Quarterly* (of the American Alliance for Health, Physical Education, and Recreation) has published only one article on the topic, and that in 1951.[10] The following is an attempt to fill this gap.

The author freely admits to a lifelong obsession with basketball as a spectator and particularly as a still-active participant in playground pickup games. His fellow players have frequently reminded him (with considerable ironic intent) that his professional knowledge of physics ought to give him an unique advantage. In truth, the major purpose of this research was to find some means to compensate for the author's stature (5'10" in sneakers), inability to leap more than 8 in. off the floor, and advancing age. With this impetus, the author has been able to demonstrate with some success that one's performance on a basketball court can be improved through the study and application of kinematics and Newtonian mechanics. Using the equations of projectile motion, a mathematical analysis of basketball shooting has been developed that can be used to determine the optimum angle and speed with which to shoot a basketball from any point on the court. But first let us consider some simple qualitative applications of physical principles to basketball play. These examples serve an additional purpose in that they can also be used to enliven the introductory physics classroom as illustrations of basic Newtonian physics.

II. WARM-UPS: PHYSICS OF SOME BASKETBALL FUNDAMENTALS

Given the interest in basketball among students at many colleges (not to mention the growth of women's basketball in recent years), illustrations of physical principles based on basketball can be used effectively in the physics lecture room as illuminating and entertaining demonstrations of how physics can be applied to seemingly unrelated situations. What follows may also be of particular interest to junior high school and high school science teachers in the inner city schools (where interest in basketball is particularly intense) as a way to motivate and interest their students.

A. Physics of the layup shot

One aspect of the principle of inertia is that if an object is initially fixed with respect to a moving reference frame and is suddenly released, the object will continue to move with the same velocity as the reference frame at the instant it is released. An early illustration of this principle was provided by Galileo in in his *Dialogues Concerning the Two Chief World Systems,* in which he pointed out that a ball released from the top of the mast of a moving ship is seen to land at the foot of the mast, and therefore must be moving horizontally as it falls with the same horizontal speed as the ship.[11]

In a basketball game, this phenomenon is illustrated by any shot taken on the run. For example, suppose that an offensive player breaks past the defense and dribbles the ball directly toward the basket as fast as he can for a layup. A few feet from the basket, he shoots the ball while still running. The inexperienced player will tend to push his shot *toward* the basket, and the result is that the ball will overshoot the rim and/or slam off the backboard with too great a speed. The player has failed to take inertia into account; by pushing his shot toward the basket he has unwittingly added to the velocity that the ball already possessed by virtue of the player's running motion. The proper way for him to take this shot is to *shoot the ball vertically upward with respect to his own body* (i.e., so the ball has no horizontal component of motion in the moving reference frame) when he is about 2–3 feet from the basket. If this is done properly, the ball should arch nicely into the basket. A

0002-9505/81/040356-10$00.50

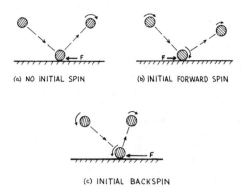

Fig. 1. Changes in translational and rotational motion of a ball when bounced on a rigid horizontal surface for various initial spin conditions.

similar situation arises when a player is moving cross court, parallel to the baseline, and takes a running shot (usually a hook shot) as he crosses the free-throw lane. If the player is moving, say, from left to right and aims his shot directly at the basket, the ball will continue its crosswise motion and will hit to the right of the center of the basket. To compensate for inertia, the player must aim the shot at the left-hand side of the rim.

B. Value of backspin

An essential principle of basketball shooting is that the ball should be pushed by the fingertips rather than by the palm of one's hand. First of all, shooting with the fingertips gives the shooter much better control over the path of the shot. Secondly, a ball shot from the fingertips and released with a slight flick of the wrist will automatically have a backspin.

Why is backspin so important? According to one of the greatest basketball coaches, Arnold "Red" Auerbach,

The fingertips ... help to impart backspin, which makes the shot softer and helps the shot to be "lucky".... A ball that strikes the rim and then stops has good backspin. Some say this is luck. But why is it that the great shooters always seem to make more of these so-called lucky shots?[12]

The value of backspin in basketball shooting is the result of good physics rather than good luck. To understand why, let us make a simple analysis of what happens when a ball bounces. We consider three cases, in which a moving ball strikes a horizontal surface with (a) no initial spin; (b) a forward spin; and (c) a backspin, as shown in Fig. 1. In each case the ball has an initial translational velocity at some angle to the surface. In what follows we shall consider just the horizontal component (parallel to the surface) of this velocity.

When the ball strikes the surface, a friction force arises at the point of contact which alters both the translational and rotational motions of the ball. In general, there will be a transfer of energy between the translational and rotational modes, with an overall net loss in the total kinetic energy. At the simplest level, it is easy to determine in each case the direction of the friction force and its effect on the motion of the ball. If the ball has no initial rotation [Fig. 1(a)] then the friction force opposes the forward translational motion. This force also creates a torque about the center of gravity of the ball, so that the ball gains angular momentum. Hence there is a loss of translational energy and an increase in

rotational energy. Thus we expect a ball that is bounced forward with no initial spin to rebound with a forward spin and a slower translational speed.

Now suppose that the ball is projected with an initial forward spin [Fig. 1(b)]. At the point of contact the rotational velocity ($v = R\omega$) will be opposite to the horizontal translational velocity component (v_x) of the ball. If the spin rate is large enough ($R\omega > v_x$) the friction force on contact will act in the forward direction. In this case, there will be an increase in the horizontal velocity and a decrease in the angular momentum; rotational energy will be converted to translational energy. The ball will appear to gain speed on the bounce, and will skip forward at a lower angle to the ground.

If the ball is thrown with a backspin [Fig. 1(c)] then the friction force at the point of contact opposes both the translational and rotational motions of the ball, and it will be larger than in the previous two cases. The result is a relatively large decrease (and possible reversal of direction) of both the translational and rotational motions. Consequently, a ball projected forward with backspin will lose considerable speed on the bounce and may even bounce backwards.

The above discussion can be presented at the most elementary classroom level as an illustration of Newton's laws of motion. Most students have played tennis, ping pong, or various wall sports (handball, paddleball, etc) at one time or another and are familiar with the way a ball behaves when it is hit with a slice (giving it a backspin) or given a topspin (forward spin). They are usually quite surprised and pleased to find that they can readily understand the physics behind these phenomena.

A more detailed quantitative analysis can be performed at the introductory college (trigonometry or calculus based) physics level. For simplicity, we assume that the bounce is elastic in the vertical direction and consider only the motion parallel to the surface. As shown in Fig. 2, the ball has an initial horizontal speed v_0 and angular velocity ω_0 (clockwise rotation is considered positive). After the bounce, the ball has a final horizontal speed and angular velocity of v_f and ω_f, respectively. It is further assumed that the ball does not skid: the point of contact comes to an instantaneous stop, so that $v_f = R\omega_f$. The equations for linear and angular momentum then yield

$$F\Delta t = \Delta(mv_x) = m(v_f - v_0),$$

$$FR\Delta t = \Delta(I\omega) = 2mR^2(\omega_0 - \omega_f)/5,$$

where Δt is the time during which the ball (radius R) is in contact with the surface. The two equations are then combined to eliminate $F\Delta t$:

$$v_f = v_0 + 2R(\omega_0 - \omega_f)/5.$$

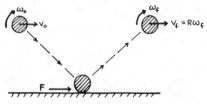

Fig. 2. Notation used in analysis of energy changes for a ball bounced on a rigid horizontal surface. It is assumed that the bounce is elastic in the vertical direction and that the ball bounces without skidding.

Table I.   Energy changes for a bouncing ball (no skidding).

	$\Delta KE_{\text{translation}}$	$\Delta KE_{\text{rotation}}$	ΔKE_{total}
No initial spin	$-\dfrac{12m}{49}v_0^2$	$\dfrac{5m}{49}v_0^2$	$-\dfrac{m}{7}v_0^2$
Initial forward spin	$-\dfrac{2m}{49}(6v_0^2 - 5R\omega_0 v_0 - R^2\omega_0^2)$	$\dfrac{m}{49}(5v_0^2 + 4R\omega_0 v_0 - 9R^2\omega_0^2)$	$\dfrac{m}{7}(v_0 - R\omega_0)^2$
Initial backspin	$-\dfrac{2m}{49}(6v_0^2 + 5R\omega_0 v_0 - R^2\omega_0^2)$	$\dfrac{m}{49}(5v_0^2 - 4R\omega_0 v_0 - 9R^2\omega_0^2)$	$\dfrac{m}{7}(v_0 + R\omega_0)^2$

The no-skidding condition $v_f = R\omega_f$ can then be substituted to eliminate either v_f or ω_f from the above equation. We are particularly interested in the energy changes that occur; i.e., the changes in the translational, rotational, and total kinetic energies of the ball. The results are as follows:

$$\Delta KE_{\text{trans}} = (1/2)mv_f^2 - (1/2)mv_0^2$$
$$= -2m(6v_0 + R\omega_0)(v_0 - R\omega_0)/49,$$

$$\Delta KE_{\text{rot}} = (1/2)I\omega_f^2 - (1/2)I\omega_0^2$$
$$= m(5v_0 + 9R\omega_0)(v_0 - R\omega_0)/49,$$

$$\Delta KE_{\text{tot}} = \Delta KE_{\text{trans}} + \Delta KE_{\text{rot}} = -m(v_0 - R\omega_0)^2/7.$$

Table I summarizes the energy changes in the three cases of (a) no initial spin; (b) initial forward spin; and (c) initial backspin. A comparison of the results clearly reveals that *a backspinning ball always experiences a greater decrease in translational energy and in total energy than a forward-spinning ball*. In addition, the energy losses suffered by a forward-spinning ball are always less than those experienced by a ball with no spin.

With these results in mind, we can now understand why a basketball launched with backspin is more likely to go in the basket. When the ball hits the rim or backboard it experiences a change in speed, spin, and energy. We have seen that a backspinning basketball always has a greater decrease in speed and energy than a forward-spinning ball does under the same conditions. This is what makes the shot seem to be "softer" and more likely to drop in the basket after it hits the rim.

A ball can also be thrown with a sideways spin. When the ball hits a surface it will veer sharply to the left or right, depending on the spin direction. Good shooters learn to make effective use of spin. One example is the reverse layup, in which a player dribbles along the baseline and under the basket. As he comes out from under the basket, he spins the ball upward against the backboard. On contact, the ball veers sharply backward into the basket.

III. BASKETBALL SHOOTING. WHAT IS THE BEST LAUNCHING ANGLE?

In this section we will use the equations of projectile motion to describe the trajectory of a basketball. The purpose of this analysis will be to determine how to take the "best" shot from any given location on the court. That is, at a given distance from the basket there are an infinite number of trajectories, each with a specific initial speed and launching angle, that connect the shooter's hand with the center of the basket. We shall attempt to identify and establish criteria to determine the "best" trajectory; i.e., the one that is most likely to result in a basket. We shall then

see whether or not the theory conforms with the real world. Since most basketball players develop their shooting ability by trial and error and constant practice, it will be interesting to find out if the "best" trajectories developed experimentally by basketball players conform to those developed theoretically.

A. Projectile motion equations

We begin by presenting the equations of motion for a body projected in a uniform gravitational field with initial speed v_0 at an angle with the horizontal of θ_0, as shown in Fig. 3. The equations for the x and y coordinates after a time t has elapsed are

$$x = v_0 t \cos\theta_0,$$
$$y = v_0 t \sin\theta_0 - (1/2)gt^2,$$

where g, the acceleration of gravity, is taken to be 32.2 ft/sec^2. The velocity components of the projectile after a time t are

$$v_x = v_0 \cos\theta_0,$$
$$v_y = v_0 \sin\theta_0 - gt.$$

The angle θ that the velocity vector of the projectile makes with the horizontal at any point on its path is given by the equation

$$\tan\theta = v_y/v_x = \tan\theta_0 - gt/v_0\cos\theta_0.$$

Through some simple manipulations of the above equations, we can eliminate the time t and arrive at the following useful relationships:

$$\tan\theta = 2y/x - \tan\theta_0, \tag{1}$$

$$v_0^2 = \frac{gx}{2\cos^2\theta_0(\tan\theta_0 - y/x)}. \tag{2}$$

The highest point of the trajectory ($y = y_{\text{max}}$) occurs when $v_y = 0$ and $\tan\theta = 0$. It can then be shown that

$$y_{\text{max}} = v_0^2 \sin^2\theta_0/2g. \tag{3}$$

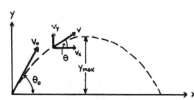

Fig. 3. Coordinate system and symbols used for description of projectile motion.

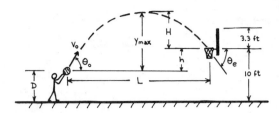

Fig. 4. Diagram and symbols used for description of trajectory of a basketball shot.

These equations describe the parabolic motion of a projectile in the absence of air resistance or other retarding forces. A typical basketball shot has a low air speed (20–30 ft/sec) and a short time of flight (~1 sec). On the other hand, a basketball is rather light in weight and has a relatively large surface area for a projectile. As a result, the effect of air resistance is small, but not negligible. Detailed calculations taking air resistance into account reveal that basketball trajectories do deviate somewhat (of the order of 5–10%) from the parabolic. Nevertheless, the assumption of parabolic motion (ignoring air resistance) does not affect the qualitative findings. We will therefore use the above equations to describe the trajectory of a basketball, and will take note afterwards of the quantitative changes produced by air resistance.

Figure 4 illustrates the path of a successful shot in the absence of air resistance. It has a trajectory which leaves the shooter's hand with an initial speed v_0 and launching angle θ_0, and passes through the point $(x = L, y = h)$, where L is the horizontal distance from the point of release to the center of the basket and h is the vertical distance between the rim of the basket and the point of release. Equation (2) becomes

$$v_0^2 = \frac{gL}{2\cos^2\theta_0(\tan\theta_0 - h/L)}.$$ (4)

In this equation, v_0 and θ_0 are the variables; for any given launching angle θ_0 within limits to be specified there is an unique positive value of v_0 that will give the desired trajectory. Equation (4) thus describes a family of parabolas that connect the point of release with the center of the basket.

Figure 5 is a graph of the relationship between v_0 and θ_0 for a typical set of values of h and L. Note that there is a minimum in the curve; in fact it can be demonstrated that there is an unique minimum initial speed for every (h,L) combination. The existence of this minimum speed v_{0m} can be shown by the standard technique of setting the first derivative $dv_0/d\theta_0 = 0$. The result obtained is that the minimum occurs at the angle θ_{0m} given by

$$\theta_{0m} = 45° + (1/2)\arctan(h/L)$$ (5a)

or

$$\tan\theta_{0m} = h/L + (1 + h^2/L^2)^{1/2}.$$ (5b)

Substitution of Eq. (5b) into Eq. (4) then yields

$$v_{0m}^2 = gL\tan\theta_{0m} = g[h + (h^2 + L^2)^{1/2}].$$ (6)

Thus of all the possible trajectories linking the point of release to the center of the basket, there is one launching angle θ_{0m} which gives the minimum-speed trajectory. The curve shown in Fig. 5 is symmetrical about a vertical axis

through θ_{0m}. Note also that in order for v_0 to be finite and real, θ_0 must be in the range $90° > \theta_0 > \arctan h/L$.

A further constraint on the allowed values of θ_0 is established by the fact that the ball must *drop* into the basket; that is, the ball must be on the descending part of its parabolic trajectory when it reaches the basket. To state this criterion mathematically, let us define the *angle of entry* θ_e as the angle between the horizontal and the tangent to the trajectory as the ball crosses the plane of the rim, as shown in Fig. 4. We can use Eq. (1) to obtain an expression for θ_e. Since θ_e is measured positively *below* the horizontal and θ is measured positively *above* the the horizontal, it follows from Eq. (1) that

$$2h/L - \tan\theta_0 = \tan\theta = \tan(-\theta_e) = -\tan\theta_e$$

or

$$\tan\theta_e = \tan\theta_0 - 2h/L.$$ (7)

Hence the condition $\tan\theta_0 > 2h/L$ must be satisfied for the ball to be on the descent when it reaches the basket. However, there is a further restriction on θ_0, as we shall now demonstrate.

B. Margin for error

According to official basketball rules, the basket must be 18 in. (1.5 ft) in diameter, and the basketball must be between $29\frac{1}{2}$ and 30 in. in circumference. Hence the diameter of a basketball is about 9.5 in. (0.79 ft), or slightly more than half the diameter of the basket. As a result, the center of the ball does not have to pass through the exact center of the basket for a score. However, it must pass through in such a way that it clears both the front and back rims. Figure 6 shows a side view of the rim; D_r is the diameter of the rim ($=1.5$ ft). $C'C$ is the path of the center of a ball that passes through the center of the basket at an angle of entry θ_e. $A'A$ is the parallel path of a ball whose center falls short of the basket center by a distance ΔL but whose edge just clears the front rim. $B'B$ is the path of a ball whose center overshoots the basket center by a horizontal distance ΔL but just misses the back rim.[13] The lines FG and AE are drawn perpendicular to $A'A$, $B'B$, and $C'C$. The ball has

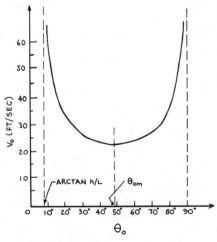

Fig. 5. Graph of relationship between v_0 and θ_0 [Eq. (4)] for a trajectory with $h = 2$ ft, $L = 13.5$ ft.

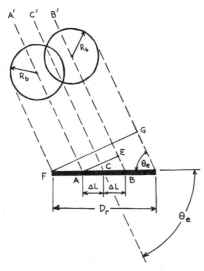

Fig. 6. Side view of basket rim showing range of successful paths of a ball passing through the basket at a given entry angle θ_e.

a diameter D_b (=0.79 ft) and radius R_b. From the geometry, we see that

$$FG = D_r \sin\theta_e,$$

$$AE = FG - 2R_b = D_r \sin\theta_e - D_b,$$

and

$$AB = 2\Delta L = AE/\sin\theta_e.$$

Combining these results we obtain

$$\Delta L = (1/2)(D_r - D_b/\sin\theta_e). \tag{8}$$

Here ΔL represents a "margin for error" in shooting; the horizontal distance L from the shooter to the basket can be changed by a maximum of $\pm\Delta L$ and the ball will still go in the basket. We note, however, that the margin for error disappears ($\Delta L = 0$) if $\sin\theta_e = D_b/D_r = (0.79)/(1.5) = 0.53$, which corresponds to $\theta_e = 32°$.

It follows that the angle of entry must be at least 32°; if θ_e is less than 32° the ball cannot clear both rims. It is possible, of course, that the ball may still bounce off the rim and/or backboard and rebound into the basket, but a shot cannot go *cleanly* through if the entry angle is less than 32°. From Eq. (7) the *minimum launching angle* θ_{0L} is therefore given by

$$\tan\theta_{0L} = \tan 32° + 2h/L = 0.62 + 2h/L. \tag{9}$$

This launching angle θ_0 must lie in the range $\theta_{0L} \leq \theta_0 < 90°$ for a successful shot. For every angle θ_0 within this range there exists a specific launching speed v_0 [given by Eq. (4)] that will make the ball pass through the center of basket. However, we have also shown that the shot will still be successful if the horizontal distance is short or long by an amount ΔL. We will now relate this allowable deviation in distance to the corresponding margins for error in speed and angle.

First, let us assume that the shooter launches the ball at the correct launching angle for a center-of-basket trajectory for the given height and distance. By what amount can his launching speed differ from the correct launching speed (v_0) such that the ball will still go in the basket? Thus for a fixed

θ_0 there is a range of values $v_0 \pm \Delta v$ that leads to a successful shot, as shown in Fig. 7. We define Δv as the *margin for error in speed*. It can be calculated directly from Eq. (4) as follows: First, v_0 is calculated for the given values of h, L, and θ_0. We replace L by $L + \Delta L$ to calculate $v_+ = v_0 + \Delta v_+$; then we replace L by $L - \Delta L$ to calculate $v_- = v_0 - \Delta v_-$. A detailed mathematical analysis shows that to a good approximation, $\Delta v_+ = \Delta v_-$. Therefore, we define $\Delta v = \Delta v_+ = \Delta v_-$.

Now let us assume that the shooter launches the ball at the correct launching speed v_0 for a center-of-basket trajectory. By what amount can the launching angle differ from θ_0 for a successful shot? There is a *margin for error in angle* $\Delta\theta_\pm$ such that if the launching angle lies between $\theta_0 - \Delta\theta_-$ and $\theta_0 + \Delta\theta_+$ the horizontal distance will fall between $L - \Delta L$ and $L + \Delta L$. Once again, Eq. (4) is used to calculate $\Delta\theta_-$ and $\Delta\theta_+$ by consecutive substitution of $L - \Delta L$ and $L + \Delta L$ for fixed values of v_0, h, and L. The behavior of $\Delta\theta_\pm$ is more complicated than that of Δv. In general, $\Delta\theta_+$ and $\Delta\theta_-$ will *not* be equal. Moreover, it can be shown that if θ_0 is less than θ_{0m} (the minimum-speed angle) then an increase in the angle causes the shot to travel a greater distance. However, if θ_0 is more than θ_{0m}, an increase in the angle makes the shot fall short, as shown in Fig. 8. Also if $\theta_0 = \theta_{0m}$, both an increase and a decrease in the launching angle will cause the shot to fall short [Fig. 8(c)].

These two margins for error, Δv and $\Delta\theta_\pm$, will serve as criteria for selecting the best trajectory. The larger the margins for error for a given shooting angle, the more freedom the shooter has to deviate from the precise values of v_0 and θ_0 needed for a center-of-basket trajectory. The aim is to find the launching angle and speed combinations that maximize the margins for error.

Thus far we have considered only the two-dimensional motion of the basketball. For a successful shot, the trajectory cannot deviate too far to the left or right of the center of the basket, or the ball will fall outside or hit the side of the rim. Consequently there is a "lateral margin for error" to be considered. This can be derived rather easily. In order for the ball to just clear the side of the rim, the center of the ball cannot deviate laterally from the center of the basket by more than $R_r - R_b$, where R_r and R_b are the radii of the rim and ball, respectively (see Fig. 9). Using the numbers given previously, we find $R_r - R_b = 0.36$ ft. This lateral displacement is over a straight-line distance from the point of release to the center of the basket of $(h^2 + L^2)^{1/2}$. Thus the *margin for error in lateral angle* $\Delta\psi$ is given by

$$\tan\Delta\psi = 0.36/(h^2 + L^2)^{1/2}. \tag{10}$$

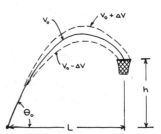

Fig. 7. Margin for error in speed: for a given launching angle θ_0, there is a range of launching speeds $v_0 \pm \Delta v$ that will result in a successful shot.

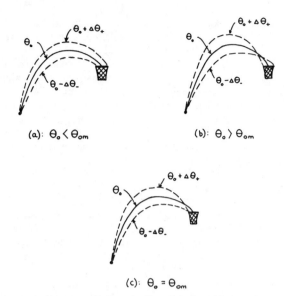

Fig. 8. Margin for error in angle: for a given launching speed v_0, any launching angle within the range $\theta_0 - \Delta\theta_-$ to $\theta_0 + \Delta\theta_+$ will result in a successful shot.

It is evident from Fig. 9 that this represents an upper limit to the margin for error on the assumption that the shot does not fall long or short of the center of the basket. The behavior of $\Delta\psi$ is rather simple because $\Delta\psi$ depends only on the launching height and distance and is independent of the launching speed and angle. Thus the closer the shooter is to the basket and the higher the launching point, the larger the margin for error and the greater the likelihood of success.

We complete the mathematical preliminaries by deriving equations for two useful quantities. The first is an estimate of the *average launching force F*. The basketball is accelerated from rest to a speed at release of v_0, so the work done on the ball is $W = (1/2)mv_0^2$, where m is the mass of the basketball (the change in gravitational potential energy during the launching of the ball can be neglected). The work done can in turn be equated to Fd, where F is the average force exerted and d is the distance the ball is moved by the shooter's hand from rest to the point of release. The distance d depends upon the player's particular shooting style. Some shooters release the ball with a sharp flick of the wrist while pivoting the forearm slightly about the elbow. Others may start the ball at the shoulder and release it when the arm is fully extended. Hence d may be between 6 in. and 3 ft. We will adopt a "typical" value of $d = 2$ ft. Also, the official weight of a basketball is 20 to 22 oz, corresponding to an average mass of 0.041 slugs. Letting $W = Fd = (1/2)mv_0^2$ and substituting numerical values, we obtain

$$F = mv_0^2/2d \approx 0.01v_0^2. \qquad (11)$$

It follows from this result that a shot launched from a given point with the minimum speed v_{0m} will require the least amount of force.

Finally, we define H, *the maximum height above the rim*:

$$H = y_{max} - h = (v_0^2 \sin^2\theta_0/2g) - h. \qquad (12)$$

This quantity will be used to determine the launching angle. When examining the trajectory of any shot, either

on sight or on film, one finds that it is rather difficult to estimate or measure the launching angle directly. However, H can be measured with little difficulty by lining up the peak of the trajectory (viewed from the side) with the plane of the rim, using the backboard as a scale of reference (see Fig. 4). For a standard rectangular (6 ft × 4 ft) backboard, the top of the backboard is 40 in. (3.3 ft) above the plane of the rim. Substituting for v_0 from Eq. (4) into Eq. (12) and solving, we obtain the following equation for θ_0 in terms of h, L, and H:

$$\tan\theta_0 = \frac{2}{L}(H + h)\left[1 + \left(\frac{H}{H + h}\right)^{1/2}\right]. \qquad (13)$$

C. Numerical results

We will now proceed to use the equations derived in Sec. III B to give numerical values for the relevant quantities. First, let us establish a reasonable range of values for the two parameters h and L. We have defined h as the vertical distance between the point of release and the rim of the basket, as shown in Fig. 4. The rim is 10 ft above the floor, and the point of release is at a height D above the floor. The value of D depends upon the height of the shooter and the type of shot taken. Generally, the ball is released at a point about 1–2 ft above the shooter's head. If the shooter is taking a jump shot, the point of release may be raised by an additional foot or so. The tallest players on occasion are actually shooting down at the basket! However, for the vast majority of players of all ages, D ranges from 6 to 9 ft. Hence h will be generally between 1 and 4 ft.

Figure 10 shows a plan of a basketball court. Most jump shots or set shots are taken at distances of 10–20 ft from the basket. A player who is closer than 10 ft to the basket is more likely to bank the shot off the backboard rather than to use a parabolic trajectory. Shots are rarely taken at distances greater than 25 ft; these are invariably desperation shots taken as time is running out. For the 1979 season, the National Basketball Association instituted a "three-point basket" for shots taken beyond a line which averages 23 ft from the basket. Only a few of these are attempted in a typical game. For our purposes, then, we will let L range from 10 to 25 ft. For the chosen ranges of h and L, the minimum-speed angle θ_{0m} lies between 45° and 55°. As the

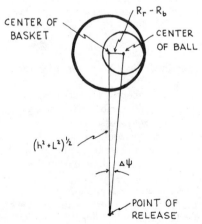

Fig. 9. Diagram for determination of margin for error in lateral angle $\Delta\psi$ (top view of basket).

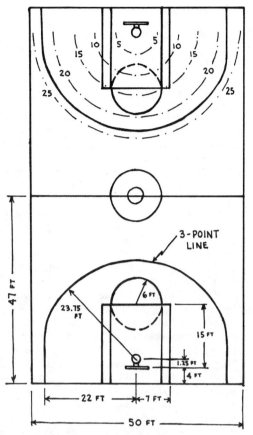

Fig. 10. Dimensions of a basketball court (professional rules). Dot–dash lines show distance from the basket in 5-ft intervals.

shooting distance and launching height increase, θ_{0m} approaches 45°.

Now let us examine the behavior of the margins for error in speed and in angle. Figure 11 shows how these quantities vary with launching angle for the specific distances $h = 2$ ft, $L = 13.5$ ft—equivalent to a shot taken from the free-throw line by a six-foot-tall player. For comparison purposes, we have plotted the *fractional* errors ($\Delta v/v_0$, $\Delta\theta_+/\theta_0$, and $\Delta\theta_-/\theta_0$). We see first of all that the margin for error in speed increases slowly as θ_0 increases and is very small, exceeding 0.01 only above 60°. This means that the shooter has very little leeway in his launching speed. That is, for a given launching angle the difference in speed between a shot that passes through the center of the basket and one that just clears either rim is generally less than 1%. To maximize the margin for error in speed, one should shoot with as high a launching angle as possible.

By comparison, the margins for error in angle tend to be much larger. Note that $\Delta\theta_+/\theta_0$ and $\Delta\theta_-/\theta_0$ each show a very sharp "peak" (actually a discontinuity) and have their largest values within a few degrees of θ_{0m}, the minimum-speed angle. To understand how these curves should be interpreted, let us take the following example. At a launching angle of 48° for the given h and L there is a specific launching speed ($v_0 = 22.46$ ft/sec) that will result in a center-of-basket trajectory. At this angle, $\Delta\theta_+/\theta_0 = 0.119$ and $\Delta\theta_-/\theta_0 = 0.038$. This means that for the same launching speed, the launching angle can be increased by 11.9% (up to 53.7°) or decreased by 3.8% (down to 46.2°)

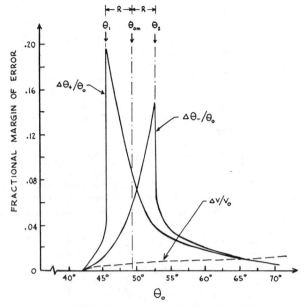

Fig. 11. Margins for error in speed and angle for trajectories with $h = 2$ ft, $L = 13.5$ ft.

and the ball will still go through the basket. As can be seen in Fig. 11, the margin for error in angle will be large if the launching angle is between θ_1 (=45.8°) and θ_2 (=52.6°), or specifically in the range 49.2 ± 3.4°, where 49.2° is the minimum-speed angle.

The behavior of the margin for error is angle near θ_{0m} can be understood by referring to Fig. 12. This shows a more detailed version of the bottom part of the $v_0 - \theta_0$ curve originally shown in Fig. 5. The solid lines contain the region of all successful shots, while the dashed line represents only the center-of-basket trajectories. It can be seen that the margin for error in angle at the particular launching speed v_1 is much larger than for any speed greater than v_1. Note that this speed corresponds to the center-of-basket trajec-

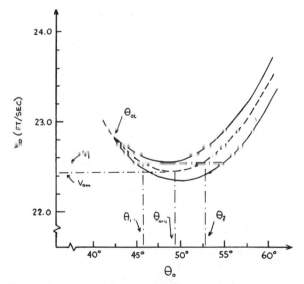

Fig. 12. Graph of v_0 vs θ_0 for $h = 2$ ft, $L = 13.5$ ft. The solid lines enclosed all points corresponding to a successful shot. The dashed line represents those points corresponding to a center-of-basket trajectory.

tory at the launching angles θ_1 and θ_2. Figure 12 demonstrates graphically why the margin for error in angle is largest in the range between θ_1 and θ_2 center on θ_{0m}. Let us define the *angular range R* as

$$R = \theta_{0m} - \theta_1 = \theta_2 - \theta_{0m}.$$

To maximize the margin for error in angle, the launching angle should therefore lie in the range $\theta_{0m} \pm R$. On the other hand, the margin for error in speed is maximized by making the launching angle as large as possible. However, $\Delta v/v_0$ varies very slowly with launching angle, while $\Delta\theta_\pm/\theta_0$ peaks very sharply near θ_{0m}. Moreover, it was noted previously that a shot launched with the minimum speed requires the smallest launching force. Taking everything into account, it seems rather clear that *the minimum-speed angle is the best launching angle*.

Several interesting aspects of basketball shooting emerge when we compare the relevant quantities for shots of different distances and heights. In Table II(a) we compare shots taken at the minimum-speed angle at a distance of 15 ft and for different values of h. The results show the advantage of a higher point of release (smaller h). As h decreases, the launching speed and force become smaller; all margins for error as well as the angular range become larger. Thus the higher the point of release, the more likely it is that the shot will be successful. This shows that given equal shooting ability, a taller player has an advantage over a shorter one *at any distance from the basket* in that the taller player has a greater margin for error. Fortunately for the shorter players, most taller players tend to develop their rebounding and under-the-basket play at the expense of their longer-distance shooting skills.

Table II(b) compares shots taken at different distances from the basket for the same launching height. As expected, the margins for error generally get smaller as the distance to the basket increases.

D. Effects of air resistance

The numerical results presented in Sec. IIIC were based on the assumption that air resistance is negligible and the trajectory is parabolic. As noted previously, air resistance *does* have a measurable effect on the trajectory of a basketball. To assess the effects of air resistance, the equations of motion for a projectile subject to a small aerodynamic drag force were calculated using the approximation methods outlined by Parker.[14] Since these calculations are too detailed to present here, we shall simply summarize the results.

The drag force on a sphere moving through air at the speeds of interest is given by the equation $F_D = (1/2) C_D \rho A v^2$, where C_D is the drag coefficient, ρ the density of air, and A the cross-sectional area of the sphere. For a basketball (weight 21 oz, diameter 0.79 ft) at speeds of 20–30 ft/sec, $C_D = 1/2$. Using $\rho = 1.37 \times 10^{-3}$ slugs/ft^3 for air, we obtain $F_D = 0.000\,29 v^2$ lb. For an air speed of 25 ft/sec, the deceleration due to drag is about 4 ft/sec^2 tangent to the path of the basketball.

A basketball launched with the speed and angle determined by the drag-free equations will typically fall short of the center of the basket by a foot or more because of the drag force. To compensate for this, the launching speed must be increased on the order of 5% for a given launching angle. The more time the ball spends in flight, the greater the effect of drag, so that more compensation is required as the launching angle and/or distance from the basket increase. In general, the minimum-speed angle will be about 1° lower than the drag-free value for a given h and L. The drag force changes the trajectory shape so as to make the descent slightly steeper than the ascent. As a result the angle of entry θ_e is increased by 2°–3°. This in turn lowers the minimum launching angle θ_{0L} by 1°–2°, and increases the margin for error in distance ΔL. The margins for error in speed and angle and the angular range R are corre-

Table II. Effect of launching height and distance on minimum speed trajectories.

| | | | | (a) Same distance to basket ($L = 15$ ft) | | | | | | |
h	θ_{0L}	θ_{0m}	R	v_{0m}	F	H	$\Delta\psi$	$\Delta v/v_{0m}$	$\Delta\theta_+/\theta_{0m}$	$\Delta\theta_-/\theta_{0m}$
1	37.0	46.9	3.9	22.72	5.16	3.28	1.37°	0.005	0.099	0.073
2	41.6	48.8	3.4	23.49	5.52	2.85	1.36°	0.004	0.085	0.058
3	45.6	50.7	2.8	24.27	5.89	2.47	1.35°	0.003	0.073	0.045
4	49.1	52.5	2:3	25.07	6.29	2.14	1.33°	0.003	0.062	0.033
				(b) Same launching height ($h = 2$ ft)						
L	θ_{0L}	θ_{0m}	R	v_{0m}	F	H	$\Delta\psi$	$\Delta v/v_{0m}$	$\Delta\theta_+/\theta_{0m}$	$\Delta\theta_-/\theta_{0m}$
10	45.6	50.7	3.5	19.82	3.93	1.65	2.02°	0.005	0.094	0.052
15	41.6	48.8	3.4	23.49	5.52	2.85	1.36°	0.004	0.085	0.058
20	39.4	47.9	3.1	26.68	7.12	4.08	1.03°	0.004	0.078	0.058
25	38.0	47.3	2.9	29.53	8.72	5.31	0.82°	0.003	0.072	0.056

h = vertical distance; point of release to rim of basket (ft).
L = horizontal distance; point of release to center of basket (ft).
θ_{0L} = minimum launching angle.
θ_{0m} = minimum speed angle.
R = angular range.
v_{0m} = minimum launching speed (ft/sec).
F = average launching force (lb).
H = maximum height above rim (ft).
Δv = margin for error in speed (ft/sec).
$\Delta\psi$ = margin for error in lateral angle.
$\Delta\theta_\pm$ = margin for error in angle.

spondingly increased. Thus the most significant effect of air resistance is that it tends to *aid* the shooter by increasing the margins for error!

IV. BASKETBALL SHOOTING: THEORY AND PRACTICE

Our development of the theory of basketball shooting has been based on the assumption that a shooter cannot always launch his shot at the precise speed and angle needed for a center-of-basket trajectory. The launching parameters can deviate somewhat from the precise values, and the shot will still be successful because of the simple fact that the ball is smaller than the basket. Our goal has been to find the conditions that maximize the margins for error, and on this basis we have concluded that the minimum-speed angle is the best launching angle. Now we must face the question as to whether this theory bears any relationship to actual basketball play.

Most basketball players are not formally taught how to shoot. Instead they learn by imitation, trial-and-error, and constant practice until they develop a "kinesthetic memory" of what they consider to be the proper launching speed and angle to be used for any given distance. The best shooters develop this ability to an extraordinary degree. We saw previously that the margin for error in speed is very small—almost always less than 1%. Thus a player who finds that his shots are falling short is able to adjust his shot by increasing the launching speed by no more than a fraction of a percent. This demonstrates the exceptional skill involved in having a good "shooting touch." The average professional player hits about 50% of his shots under game conditions; for open (unguarded) shots or in practice the average increases to 70% or more. When shooting a free throw (foul shot) the player is allowed time to relax and concentrate on his shot. Even allowing for shots that hit the rim or backboard and rebound into the basket, the margins for error are still rather small. Yet the best professionals hit this shot 90% of the time![15] It is rather remarkable that the human body can be trained to reproduce the required movements so precisely.

What launching angles do most basketball players prefer to use? The author's observations on this matter are based on countless hours of playing, watching, and coaching basketball, as well as on some experimentation with his own shooting. First, let us consider the "high-arch" shot—one that is launched at an angle well above the minimum-speed angle. The high-arch shot in theory has the advantage of a larger margin for error in speed. It is also more difficult to block (it can be launched over the outstretched hand of a taller defender). However, in practice it is a rather difficult shot to launch and to aim. For example, a shot taken by a six footer ($h = 2$ ft) from the free-throw line and launched at a 65° angle would have a maximum height H above the rim of 5.77 ft, or 2.44 ft above the top of the backboard. In fact, one rarely ever sees a shot from this distance go higher than the top of the backboard ($H = 3.33$ ft). Using Eq. (13) we find that this corresponds to a launching angle of 50° to 60°, depending on the height of the shooter. As a general rule, in actual play no shot is ever launched at an angle greater than 60° in the 10–25-ft shooting range.

At the other extreme is the "flat" or low-arch shot, which is aimed almost on a line at the basket, and is launched at an angle well below the minimum-speed angle. (It may even be launched at an angle below θ_{0L}, so that the entry angle is less than 32° and a good bounce is needed for the shot to be successful.) This type of trajectory is commonly seen in playground basketball. It is used especially by youngsters who, because of a lack of height and strength, have developed a habit of aiming the ball *at* the rim instead of trying to arch the shot *over* the rim. The theory shows, however, that a flat shot has smaller margins for error and actually requires more launching force in comparison to a shot taken with a higher arch (closer to the minimum-speed angle).

The style of play in many ways determines the type of shot to use. In the early days of basketball, team play was slower and more deliberate. Players tended to run in fixed patterns and the ball was passed frequently until one player managed to get free for an open shot. The most common type of shot was the two-handed set shot, a relatively high-arch shot launched from the chest or over the head with both feet on the floor. Since the late 1950s the style of play has changed noticeably. There is now a greater emphasis on speed and individual "one-on-one" play. The most common shot is the jump shot, which is launched (usually one-handed) while in midair. Often, a player will dribble to a specific spot on the floor, come to a sudden stop, leap vertically or slightly backwards, and release his shot. A jump shot must be launched quickly with a minimum of effort. Obviously, the minimum-force feature of the minimum-speed angle is a big advantage here. The jump shot has become so popular that many players use it even when they are unguarded, in preference to a set shot taken with both feet on the floor (presumably giving the shooter better balance and control). We have seen that there is a theoretical advantage to this, because the margins for error increase as the launching height increases [see Table II(a)]. However, there is a tendency to launch a jump shot on a flat trajectory. This is a disadvantage to the shooter, in that a slightly higher-arch shot can be launched with less force and better accuracy.[16] There is also a tendency to launch all shots at the same angle regardless of distance. However, the minimum-speed angle increases as the distance decreases [see Table II(b)]. Therefore, one should learn to shoot at a higher angle as one gets closer to the basket.

As a simple test of the theory, a brief study was made to determine the launching angles used by good shooters. A number of Brooklyn College students (all of them considered to be good playground ballplayers) were filmed using a super-8 movie camera as they took set and jump shots at various distances from the basket. The developed film was then run at slow speed through a film editor viewer, and the trajectory of each shot was traced out on a transparent plastic sheet placed over the viewer screen. The launching angle could then be calculated with reasonable accuracy from the shape and maximum height of the trajectory. Nearly 80 shots were analyzed in this fashion. The results showed that the successful shots were being launched at angles very close to the minimum-speed angle and within the angular range for the given height and distance. These very limited and relatively crude measurements seem to indicate that the better shooters do learn to shoot at or near the minimum-speed angle.

A more detailed study of preferred shooting angles might be of particular interest to basketball coaches as well as to physical education faculty in the areas of motor learning, kinesiology, and biomechanics. Results could be used to improve the teaching of basketball shooting to youngsters. One technique suggested by Mortimer[10] is to suspend strings or other markers at the appropriate height and distance of the trajectory peak to be used as aiming targets.

V. SUMMARY AND CONCLUSIONS

The major purpose of this work has been to show how a knowledge of physics can be used to understand certain aspects of basketball play and to improve one's performance as a player. We have seen how an awareness of the principle of inertia and an understanding of the effects of spin can help to improve specific shooting techniques. Our analysis of the trajectory of a basketball has demonstrated the existence of a best shooting angle—namely, the minimum-speed angle. There are clear advantages to shooting at this angle in comparison to the more commonly used flat trajectory.

Thus a knowledge of physics *can* make one a better basketball player. The results obtained here should be useful in teaching youngsters how to play the game. An additional benefit is that many of the examples used in this work can be presented in the physics classroom as interesting illustrations of basic Newtonian principles.

[1] L. J. Briggs, Am. J. Phys. **27**, 589 (1959).

[2] P. Kirkpatrick, Am. J. Phys. **31**, 606 (1963).

[3] S. Chapman, Am. J. Phys. **36**, 868 (1968).

[4] R. G. Watts and E. Sawyer, Am. J. Phys. **43**, 960 (1975).

[5] H. Brody, Am. J. Phys. **47**, 482 (1979).

[6] D. C. Hopkins and J. D. Patterson, Am. J. Phys. **45**, 263 (1977).

[7] T. Jorgenson, Am. J. Phys. **38**, 644 (1970).

[8] C. Frohlich, Am. J. Phys. **47**, 583 (1979).

[9] J. D. Memory, Am. J. Phys. **41**, 1205 (1973).

[10] E. M. Mortimer, Res. Q. **22**, 234 (1951).

[11] G. Galilei, *Dialogues Concerning the Two Chief World Systems,* translated by Stillman Drake, 2nd ed. (University of California, Berkeley and Los Angeles, 1967), p. 126 and 144.

[12] A. Auerbach, *Basketball for the Player, the Fan, and the Coach* (Pocket, New York, 1976), p. 80.

[13] A less stringent criterion would be that the center of the ball (rather than the edge) must pass inside the edge of the rim, since the ball would then hit the rim and probably drop into the basket. This would allow a somewhat larger margin for error. However, we shall adopt the purist approach accepted by the best shooters, who like to make the ball "swish" through the basket without touching the rim.

[14] G. W. Parker, Am. J. Phys. **45**, 606 (1977).

[15] According to the *Guinness Book of World Records* (Bantam, New York, 1980), the record for consecutive free throws is 2036, set by Ted St. Martin in 1977. In 1978, Fred L. Newman made 88 consecutive free throws while blindfolded!

[16] The author's own experience serves to verify this conclusion. For many years he used a flat shot, with consistently mediocre results. A shift in shooting technique to a higher launching angle led, after only a few weeks of practice, to a significant and very satisfying improvement in shooting accuracy.

R. L. Huston

Professor of Mechanics,
Department of Engineering Science,
University of Cincinnati,
Cincinnati, Ohio 45221
Mem. ASME

C. Passerello

Associate Professor of Engineering Mechanics,
Department of Mechanical Engineering and
Engineering Mechanics,
Mechanical Technological University,
Houghton, Mich. 49931

J. M. Winget

Graduate Student,
Department of Applied Mechanics,
California Institute of Technology,
Pasadena, Calif. 91125

J. Sears

Research Assistant,
Department of Engineering Science,
University of Cincinnati,
Cincinnati, Ohio 45221

On the Dynamics of a Weighted Bowling Ball[1]

An analysis of the dynamics and performance of a weighted, slipping/rolling bowling ball is presented. The analysis uses Euler parameters and angular velocity components as dependent variables. The governing equations of motion are integrated using standard digital/numerical procedures. Particular attention is given to factors affecting ball performance ("hook") and the lane oil tracing on the ball. It is found that factors most affecting hook are the mass-center location, the lane conditions (friction), and the initial angular velocity component parallel to the lane.

Introduction

Recently, sports, physical fitness, and related leisure-time activities have occupied the attention, time, and energies of people of virtually every age, locale, and vocation. This surge of popularity of leisure-time activities has in turn, stimulated analysts and engineers to seek quantitative descriptions of these activities with the general objectives of increasing conceptual understanding, performance, and safety. Moreover, the efforts of these analysts have been recently enhanced with the development of new analytical procedures and techniques, made possible primarily through the advent of high-speed digital computers. This paper presents the results of applying some of these new techniques in a study of the dynamics of a bowling ball. Specifically, the recently exposited [1] technique of using Euler parameters to avoid kinematical singularities is employed to describe the instantaneous orientation of the sliding, rotating ball. This together with a computer oriented formulation of the governing equations of motion form the basis of the analysis.

The specific objectives of this study are fourfold:

1 To determine the effect of the various components of the initial angular velocity of the ball on the performance, that is, the hook of the ball.

2 To determine the effect of mass center offset on the hook.

3 To examine the pattern of "ball-track" on the ball under various conditions.

4 To develop a user-oriented computer code to help provide knowledge and insight for competitive bowlers, ball manufacturers, and ball drillers of the quantitative as well as qualitative effect of factors such as mass-center offset, lane smoothness, and initial position, velocity, and angular velocity on the subsequent kinematics of the ball.

The balance of the paper is divided into five parts with the next three parts containing the elements of the notation, geometry, kinematics, kinetics, and dynamics of the ball and bowling lane. This is followed by a presentation of the results of ball performance under a variety of different initial configurations and conditions. The final part contains conclusions and observations about the analysis and its applications.

Preliminary Considerations

Notation. For the purposes of the analysis let the ball and bowling lane be represented as shown in Fig. 1, where B is the bowling ball,

[1] This work was supported in part by the National Science Foundation under Grant ENG 75 21037.

Contributed by the Applied Mechanics Division for presentation at the Winter Annual Meeting, New York, N. Y., December 2–7, 1979, of THE AMERICAN SOCIETY OF MECHANICAL ENGINEERS.

Discussion on this paper should be addressed to the Editorial Department, ASME, United Engineering Center, 345 East 47th Street, New York, N. Y. 10017, and will be accepted until two months after final publication of the paper itself in the JOURNAL OF APPLIED MECHANICS. Manuscript received by ASME Applied Mechanics Division, December, 1978.   Paper No. 79-WA/APM-17.

Copies available until September, 1980.

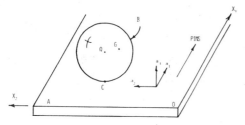

Fig. 1   Bowling ball and lane

A is the lane, Q is the geometric center of B, G is the mass center of B, and C is that point of B which in a given instant is in contact with A. Fig. 1 also depicts three unit vectors a_1, a_2, and a_3 whose orientation are fixed in A. a_3 is vertically up and thus is normal to the plane of A. a_1 is directed "down the alley," that is, toward the pins, and a_2 is generated by the vector product of a_3 and a_1 and directed to the left of the lane. Finally, Fig. 1, contains coordinate axes X_1 and X_2 originating at 0, the right corner of A.

Coordinates and Degrees of Freedom. Throughout the analysis it is assumed that the ball B either slips or rolls relative to the lane A and that at any instant there is always one, and only one point C of B which is in contact with A. If B slips on A (which is usually the case in bowling), it has 5 degrees of freedom: 2 in translation and 3 in rotation. If B rolls on A (as occassionally occurs near the end of a bowling "shot"), it has only 3 degrees of freedom since, by the definition of rolling, the contact point C then has zero velocity in A. (That is, the translation and rotation of B are no longer independent.)

The translation of B may be described by the X_1 and X_2 coordinates of Q. The rotation or orientation of B may be described in terms of four coordinates $\epsilon_i (i = 1, \ldots, 4)$ called "Euler parameters" (of which three are independent). These parameters may be defined by first recalling that B may be brought into any given orientation relative to A by a rotation about an appropriate axis [2]. If λ is a unit vector along this axis and if θ is the rotation angle, the ϵ_i are [1–3]

$$\epsilon_1 = \lambda_1 \sin \theta/2$$

$$\epsilon_2 = \lambda_2 \sin \theta/2 \qquad (1)$$

$$\epsilon_3 = \lambda_3 \sin \theta/2$$

$$\epsilon_4 = \cos \theta/2$$

where λ_1, λ_2, and λ_3 are the a_1, a_2, and a_3 components λ. From equations (1) it is seen that

$$\epsilon_1{}^2 + \epsilon_2{}^2 + \epsilon_3{}^2 + \epsilon_4{}^2 = 1 \qquad (2)$$

Singularities and Euler Parameters. As mentioned earlier, the reason for using Euler parameters instead of the more commonly used orientation angles (for example, Euler angles), is that the Euler parameters do not introduce the singularities associated with orientation angles [1, 3]. For example, if orientation angles are used to describe the orientation of B relative to A, it is easily shown that the a_1, a_2, and a_3 components of the angular velocity of B relative to A may be expressed as [4–6]

$$\omega_1 = \dot{\alpha} + \dot{\gamma} S_\beta$$

$$\omega_2 = \dot{\beta} C_\alpha - \dot{\gamma} S_\alpha C_\beta \qquad (3)$$

$$\omega_3 = \dot{\beta} S_\alpha + \dot{\gamma} C_\alpha C_\beta$$

where S and C are abbreviations for sine and cosine and α, β, and γ are the dextral orientation angles defined as follows: Imagine a set of unit vectors b_1, b_2, and b_3 fixed in B and aligned (initially) with a_1, a_2, and a_3. Then B may be brought into general orientation relative to A by three successive dextral rotations of B about b_1, b_2, and b_3

through the angles α, β, and γ. If equations (3) are solved for α, β, and γ in terms of ω_1, ω_2, and ω_3 (a process similar to that occurring in the numerical solution of governing equations of motion), one obtains

$$\dot{\alpha} = [\omega_1 C_\beta + S_\beta(\omega_2 S_\alpha - \omega_3 c_\alpha)]/C_\beta$$

$$\dot{\beta} = [\omega_2 C_\alpha C_\beta + \omega_3 S_\alpha C_\beta]/C_\beta \qquad (4)$$

$$\dot{\gamma} = [\omega_3 C_\alpha - \omega_2 S_\alpha]/C_\beta$$

A singularity is thus found when $\beta = 90°$. (A similar singularity occurs with other choices of orientation angles [3].) It is shown below that this kind of singularity is avoided entirely by using the Euler parameters as defined in equations (1). However, the price paid for this convenience is the introduction of an additional variable and a corresponding constraint equation (2).

When using Euler parameters the angular velocity components (analogous to equations (3)) take the form [1–3]

$$\omega_1 = 2(\epsilon_4 \dot{\epsilon}_1 - \epsilon_3 \dot{\epsilon}_2 + \epsilon_2 \dot{\epsilon}_3 - \epsilon_1 \dot{\epsilon}_4)$$

$$\omega_2 = 2(\epsilon_3 \dot{\epsilon}_1 + \epsilon_4 \dot{\epsilon}_2 - \epsilon_1 \dot{\epsilon}_3 - \epsilon_2 \dot{\epsilon}_4) \qquad (5)$$

$$\omega_3 = 2(-\epsilon_2 \dot{\epsilon}_1 + \epsilon_1 \dot{\epsilon}_2 + \epsilon_4 \dot{\epsilon}_3 - \epsilon_3 \dot{\epsilon}_4)$$

Solving for the $\epsilon_i (i = 1, \ldots, 4)$ in terms of the $\omega_i (i = 1, 2, 3)$, one obtains

$$\dot{\epsilon}_1 = \tfrac{1}{2}(\epsilon_4 \omega_1 + \epsilon_3 \omega_2 - \epsilon_2 \omega_3)$$

$$\dot{\epsilon}_2 = \tfrac{1}{2}(-\epsilon_3 \omega_1 + \epsilon_4 \omega_2 + \epsilon_1 \omega_3) \qquad (6)$$

$$\dot{\epsilon}_3 = \tfrac{1}{2}(\epsilon_2 \omega_1 - \epsilon_1 \omega_2 + \epsilon_4 \omega_3)$$

$$\dot{\epsilon}_4 = \tfrac{1}{2}(-\epsilon_1 \omega_1 - \epsilon_2 \omega_2 - \epsilon_3 \omega_3)$$

(This solution is quickly obtained by observing that if equation (2) is differentiated and annexed to equations (6), the set of equations could be written in the matrix form

$$\begin{bmatrix} \omega_1 \\ \omega_2 \\ \omega_3 \\ \omega_4 \end{bmatrix} = 2 \begin{bmatrix} \epsilon_4 & -\epsilon_3 & \epsilon_2 & -\epsilon_1 \\ \epsilon_3 & \epsilon_4 & -\epsilon_1 & -\epsilon_2 \\ -\epsilon_2 & \epsilon_1 & \epsilon_4 & -\epsilon_3 \\ \epsilon_1 & \epsilon_2 & \epsilon_3 & \epsilon_4 \end{bmatrix} \begin{bmatrix} \dot{\epsilon}_1 \\ \dot{\epsilon}_2 \\ \dot{\epsilon}_3 \\ \dot{\epsilon}_4 \end{bmatrix} \qquad (7)$$

where ω_4 is equal to the derivative of equation (2) and has the value zero. The square matrix is seen to be orthogonal (that is, the transpose is the inverse) and hence equations (6) follow immediately from (7) upon letting ω_4 be zero). Equations (6) which are analogous to equations (4) thus do not contain a singularity.

In the analysis which follows, it is convenient to express the physical, kinematical, and kinetic vectors in terms of the unit vectors $a_i (i = 1, 2, 3)$ fixed in the lane A. To do this, it is useful to introduce an orthogonal transformation matrix called a "shifter" [4, 5, 7] from the ball to the lane with elements defined as

$$S_{ij} = a_i \cdot b_j \qquad (8)$$

where, as before, b_i ($i = 1, 2, 3$) are a dextral set of mutually perpendicular unit vectors fixed in the ball B, with b_2 directed into the finger holes, b_3 is directed toward and perpendicular to the thumb hole, and b_1 is generated by the vector product of b_2 and b_3. (b_1, b_2, and b_3 are thus approximately aligned with a_1, a_2, and a_3 at the start of the "shot.") Then, for example, if a vector V is expressed in terms of the b_i, S_{ij} may be used to express V in terms of the a_i through the relations

$$V_{ai} = S_{ij} V_{bj} \qquad (9)$$

where V_{ai} and V_{bi} are the a_i and b_i components of V and where a repeated index (that is, j) represents a sum from 1 to 3 over that index. Finally, through using equations (1) and (8), S_{ij} may be expressed in terms of the Euler parameters ϵ_i ($i = 1, \ldots, 4$) as [2, 3]:

FULL ROLLER   SEMI ROLLER   SPINNER

Fig. 2 Typical ball-tracks

ball-track of considerably smaller radius than the ball radius as also shown in Fig. 2.

To describe the ball-track in terms of the ball's kinematics, consider the following: The translation of Q, the ball center may be ignored in discussions about the ball track. That is, a complete description of the ball track may be obtained by an analysis of the ball *rotation* relative to the bowling lane A. Hence, imagine A to be covered with an ink or dye which traces out a curve on the ball as depicted in Fig.

$$S_{ij} = \begin{vmatrix} \epsilon_1{}^2 - \epsilon_2{}^2 - \epsilon_3{}^2 - \epsilon_4{}^2 & 2(\epsilon_1\epsilon_2 - \epsilon_3\epsilon_4) & 2(\epsilon_1\epsilon_3 + \epsilon_2\epsilon_4) \\ 2(\epsilon_1\epsilon_2 + \epsilon_3\epsilon_4) & -\epsilon_1{}^2 + \epsilon_2{}^2 - \epsilon_3{}^2 + \epsilon_4{}^2 & 2(\epsilon_2\epsilon_3 - \epsilon_1\epsilon_3) \\ 2(\epsilon_1\epsilon_3 + \epsilon_2\epsilon_4) & 2(\epsilon_2\epsilon_3 + \epsilon_1\epsilon_4) & -\epsilon_1{}^2 - \epsilon_2{}^2 + \epsilon_3{}^2 + \epsilon_4{}^2 \end{vmatrix} \quad (10)$$

Kinematics

Angular Velocity and Angular Acceleration. Let the angular velocity of the ball B relative to the lane A be expressed as

$$\omega = \omega_1\mathbf{a}_1 + \omega_2\mathbf{a}_2 + \omega_3\mathbf{a}_3 \quad (11)$$

Then the ω_i $(= 1, 2, 3)$ are functions of ϵ_i and ϵ_i $(i = 1, \ldots, 4)$ as given by equations (5). Now, since the $\mathbf{a}_i$ $(i = 1, 2, 3)$ are fixed in A, the angular acceleration of B in A is simply [6]

$$\alpha = \dot\omega_1\mathbf{a}_1 + \dot\omega_2\mathbf{a}_2 + \dot\omega_3\mathbf{a}_3 \quad (12)$$

In the sequel, the ω_i become the dependent variables of the governing equations of motion. They are a measure of the *rate of change* of orientation of B relative to A. Once they are known, the orientation itself, as defined by the ϵ_i, may be determined from equations (6).

Velocity and Acceleration. The position of the ball center Q relative to the lane A is defined by the coordinates x_1, x_2, and x_3 with x_3 being the (constant) ball radius r. Hence, the velocity and acceleration of Q relative to A are

$$\mathbf{V}_Q = \dot{x}_1\mathbf{a}_1 + \dot{x}_2\mathbf{a}_2 \quad (13)$$

and

$$\mathbf{a}_Q = \ddot{x}_1\mathbf{a}_1 + \ddot{x}_2\mathbf{a}_2 \quad (14)$$

Now, since the mass-center G, like Q, is fixed in B, its velocity and acceleration relative to A are [6]

$$\mathbf{V}_G = \mathbf{V}_Q + \omega \times \mathbf{P} \quad (15)$$

and

$$\mathbf{a}_G = \mathbf{a}_Q + \alpha \times \mathbf{P} + \omega \times (\omega \times \mathbf{P}) \quad (16)$$

where $\mathbf{P}$ is the position vector (fixed in $\mathbf{B}$) locating G relative to Q. (In the computation of the vector products in equations (15) and (16), it is convenient to express $\mathbf{P}$ in terms of the a_i $(i = 1, 2, 3)$ by using equation (9).)

Finally, since C is also fixed in B, its velocity relative to A is

$$\mathbf{V}_C = \mathbf{V}_Q + \omega \times (-r\mathbf{a}_3) \quad (17)$$

when B rolls on A, $\mathbf{V}_C$ is zero. Then, by equations (11), (13), and (17) $\dot{x}_1$ and $\dot{x}_2$ become

$$\dot{x}_1 = r\omega_2 \quad \text{and} \quad \dot{x}_2 = -r\omega_1 \quad (18)$$

Ball Track. Recall that an objective of the analysis is to determine the relationship between the "ball-track" and the kinematics of the ball. The ball-track is the trace of oil and dirt from the bowling lane which accumulates on the ball. The ball-track is of interest to bowlers since it is said to be a measure of the form and technique of the ball "release" at the start of the shot [8, 9]. Specifically, a bowler is said to have "good form and control" if the ball-track is a "semiroller"— that is, a circle of slightly smaller radius than the ball radius and passing beneath the thumb and finger holes as shown in Fig. 2. A "full-roller" is a ball-track along a great circle, and a "spinner" is a

3. The rate of curve tracing is proportional to the angular speed of the ball and the distance ξ between the contact point C and a diametral line which is instantaneously parallel to the angular velocity vector. The instantaneous orientation of the curve tracing is perpendicular to this "angular velocity diameter." If the angular velocity ω is a constant, the ball-track is a circle with radius ρ given by

$$\rho = r \sin \psi = r(\omega_1{}^2 + \omega_2{}^2)^{1/2}/|\omega| \quad (19)$$

where ψ is the angle between ω and the normal to A, as shown in Fig. 3 and $|\omega|$ is the magnitude of ω. (ψ is simply a measure of the inclination of ω and it may be expressed as $\cos^{-1}(\omega_3/|\omega|)$.) If ω is not constant the ball-track is not, in general, a circle but it is instead a curve whose instantaneous radius of curvature is ρ as given in equation (19). Equation (19) shows, however, that if $|\omega|$ and ω_3 are both constant, the ball-track is still a circle. This means that the ball-track is independent of the direction of the projection of ω on A. Finally, if $|\omega|$ is nearly a constant and the change in ω_3 during a shot is small (as is usually the case), then ψ is nearly a constant and the ball track is approximately circular with radius $r \cos \psi$. In this case the ball-track is contained within a circular band of width w given approximately by

$$w = r\Delta\psi = r(\omega_1{}^2 + \omega_2{}^2)^{1/2}\Delta\epsilon_3/|\omega|^2 \quad (20)$$

where $\Delta\psi$ and $\Delta\omega_3$ are the changes in ψ and ω_3 during the shot. (This expression is obtained by differentiating equation (19) while holding $(\omega_1{}^2 + \omega_2{}^2)^{1/2}$ nearly constant.)

A numerical representation of ball-track may be obtained by keeping a tabular record of the coordinates of the contact points of

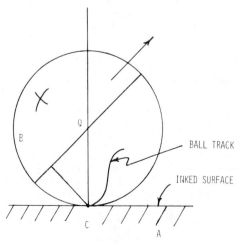

Fig. 3 Ball-track from ball rotation on an ink-covered surface

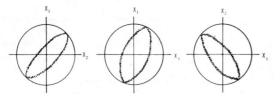

Fig. 4   Typical ball-track from computer simulation

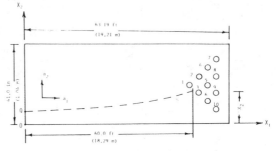

Fig. 5   Bowling lane and path of a typical shot

the ball during the course of the shot. Specifically, this may be done by expressing the position vector locating C relative to Q in terms of the unit vectors $\mathbf{b}_i (i = 1, 2, 3)$ fixed in B. Hence, by equation (9), this becomes

$$-r\mathbf{a}_3 = -r(S_{31}\mathbf{b}_1 + S_{32}\mathbf{b}_2 + S_{33}\mathbf{b}_3) \qquad (21)$$

Then a plot of the coordinates $-r[S_{31}, S_{32}, S_{33}]$ is a representation of the ball-track relative to the ball itself. Three planar views of such a plot for a typical computer simulation (as described later in the paper) of a bowling shot are shown in Fig. 4.

Equations of Motion

Perhaps the most direct method of obtaining governing equations of motion for this system is to examine a free-body diagram of the ball and to set resultant forces and moments zero by d'Alembert's principle [10].

Kinetics. In such a free-body diagram, the externally applied or *active* forces are quivalent to a gravitational force $-mg\,\mathbf{a}_3$, passing through the mass-center G, and a contact force $\mathbf{C}$ passing through the contact point C, where m is the mass of the ball B and g is the gravitational constant. When B slips on A, the contact force $\mathbf{C}$ may be written as

$$\mathbf{C} = -\mu N\tau + N\mathbf{a}_3 \qquad (22)$$

where N is the magnitude of the resultant vertical or normal force exerted by the lane A on B, μ is the coefficient of friction, and τ is a unit vector in the direction of the velocity of C relative to A.

The *inertia* forces exerted on B are equivalent to a single force $\mathbf{F}$, passing through G, together with a couple with torque $\mathbf{T}$, where $\mathbf{F}$ and $\mathbf{T}$ are [6, 10]

$$\mathbf{F} = -m\,\mathbf{a}_G \qquad (23)$$

and

$$\mathbf{T} = -\mathbf{I} \cdot \alpha - \omega \times \mathbf{I} \cdot \omega \qquad (24)$$

where $\mathbf{I}$ is the inertia dyadic of B relative to G.

Governing Equations. Consider first the case when B slips on A: Using d'Alembert's principle and setting the resulting of the active and inertia forces exerted on B equal to zero leads to the vector equation

$$\mathbf{F} + \mathbf{C} - mg\,\mathbf{a}_3 = 0 \qquad (25)$$

Also, setting the resultant moment about Q, the ball center, equal to zero, leads to

$$\mathbf{T} - \mathbf{P} \times m(g\,\mathbf{a}_3 + \mathbf{a}_G) - r\mathbf{a}_3 \times \mathbf{C} = 0 \qquad (26)$$

These two vector equations are equivalent to six scalar equations which may be obtained by using equations (11), (12), (16), (22)–(24), and setting the respective $\mathbf{a}_i$ $(i = 1, 2, 3)$ components equal to zero. The unknown dependent variables appearing in these six equations are: x_1 and x_2, the translation coordinates of Q; ω_i $(i = 1, 2, 3)$, the angular velocity components of B; and N, the normal force exerted by A on B. N is easily eliminated from these equations, and if this is done the resulting five equations may be written in the form

$$a_{ij}\dot{u}_j = f_i \qquad (27)$$

where the indices i and j range from 1–5 and there is a sum over j. a_{ij} and f_i are tabulated in the Appendix and the u_j are: $u_1 = \dot{x}_1$, $u_2 = \dot{x}_2$, $u_3 = \omega_1$, $u_4 = \omega_2$, and $u_5 = \omega_3$.

When B rolls on A, the equations of motion may be obtained by simply setting the resultant moment about C equal to zero. That is,

$$\mathbf{T} - (r\mathbf{a}_3 + \mathbf{P}) \times m(g\,\mathbf{a}_3 + \mathbf{a}_G) = 0 \qquad (28)$$

Following the same procedure as just outlined, this equation is found to be equivalent to three scalar equations which may be written in the form

$$b_{ij}\dot{\omega}_j = g_i \qquad (29)$$

where the indices i and j range from 1–3 and there is a sum over j. b_{ij} and g_i are also tabulated in the Appendix.

In a typical bowling shot B slips on A for almost the rull range of the shot with rolling possibly occuring near the end of the shot. Therefore, the governing equations are (27) until or unless rolling occurs. The condition required for rolling to occur is that $\mathbf{V}_C$, given by equation (17), be zero. In this case, the governing equations are then (29).

If B is not homogeneous, it is possible, under conditions of varying lane friction coefficient, for slipping to recur after rolling. This would happen if the magnitude of the component of F parallel to A, exceeds μN. The governing equations would then revert to (27).

Numerical Solutions. Except for the case when B is homogeneous, closed-form solutions of equations (27) and (29) are not generally available. Hence, numerical solutions are sought as an alternative. In the case when both slipping and rolling could occur during the course of a shot, it is necessary to solve equations (27) and (29) in tandem. Moreover, the magnitudes of $\mathbf{V}_C$ and $\mathbf{F}$ need to be continually evaluated to determine which set of governing equations is valid.

The components of $\mathbf{P}$ and $\mathbf{I}$ occurring in the coefficients a_{ij}, f_i, b_{ij}, and g_i of equations (27) and (29) are referred to $\mathbf{a}_i$ $(i = 1, 2, 3)$, the unit vectors fixed in A. This means they are not constants but are instead functions of the four Euler parameters $\epsilon_i (i = 1, \ldots, 4)$ through the shifter transformation matrices introduced and defined in equations (8) and (10).[2] Therefore, equations (6) must be adjoined to equations (27) and (29). For the slipping case, this leads to a total of nine first-order equations for the nine dependent variables u_i $(i = 1, \ldots, 5)$ and ϵ_i $(i = 1, \ldots, 4)$. For rolling, this results in seven first-order equations for the seven dependent variables ω_i $(i = 1, 2, 3)$ and ϵ_i $(i = 1, \ldots, 4)$. Given suitable initial conditions, these sets of equations may be solved numerically and in tandem using one of the standard integration routines.[3] In the following part of the paper the results of such nu-

---

[2] Note that if $\mathbf{P}$ and $\mathbf{T}$ are referred to $\mathbf{b}_i (i = 1, 2, 3)$, the unit vectors fixed in B, their components are constants. This suggests that it might be more convenient to develop the governing equations in terms of $\mathbf{b}_i$ instead of $\mathbf{a}_i$. However, it is quickly seen that this approach introduces Euler parameters into the governing equations through the transformation of the components of the gravitational and contact forces.

[3] A tape containing a source listing, sample runs, and a users manual of a computer program developed for this purpose is available from the authors at reproduction cost.

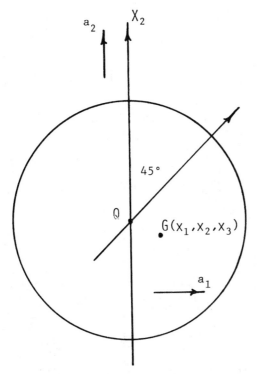

Fig. 6   Initial angular velocity for computer runs with displaced mass-center; see Table 2

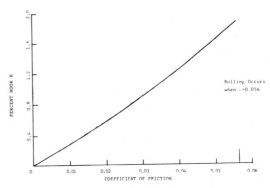

Fig. 7   Relation between hook and lane friction

Table 1   Effect of initial angular velocity direction on hook

INITIAL ANGULAR VELOCITY (RAD/SEC)			PERCENT HOOK
ω_1	ω_2	ω_3	H
0.0	10.0	0.0	0.0
-2.588	9.659	0.0	0.34
-5.0	8.66	0.0	0.64
-7.07	7.07	0.0	0.87
-9.659	2.588	0.0	1.10
-10.0	0.0	0.0	1.09
-6.96	6.96	1.736	0.86
-6.64	6.64	3.42	0.82
-6.12	6.12	5.0	0.75
-7.07	7.07	0.25	0.87
-7.07	7.07	0.50	0.87
-7.07	7.07	1.0	0.87
-7.07	7.07	2.0	0.87
-7.07	7.07	5.0	0.87

merical solutions using a fourth-order Runge-Kutta technique are presented for a variety of initial configurations and physical conditions.

Numerical Results

Hook. Fig. 5 depicts a top view (not to scale) of a bowling lane and a typical shot. As mentioned earlier, a measure of the "performance" of the ball is its "hook," that is, its x_2 displacement Δx_2, at the end of the shot. For computational and graphical purposes it is convenient to define the "hook percentage" H as the ratio (in percent) of Δx_2 and d, the distance from the foul line to the head pin. Hence,

$$H = 100\, \Delta x_2/18.288 \qquad (30)$$

where Δx_2 is measured in meters.

Numerical Simulations. Three series of numerical simulations were made to determine the factors affecting H. This involved numerically integrating equations (6) and (25) for a ball with a 0.109m (0.358 ft) radius and weighing 71.32N (16.0 lb). In the first series, H was measured while the direction of the initial angular velocity was varied, with the magnitude held constant at 10 rad/sec for a ball with its mass center G at the geometric center Q($\mathbf{P} = 0$). In the second series, H was measured while the position of G was varied on a sphere of radius 0.0015m (0.005 ft) about Q. In this series the initial angular velocity was kept constant in *both* magnitude and direction with the magnitude being 10 rad/sec and the direction being parallel to the lane surface A and inclined at 45° to $\mathbf{a}_1$ as shown in Fig. 6. A third series of runs was made under the same condition as the second series except that inertia components of the ball were varied while the mass center G remained at the geometric center C ($\mathbf{P} = 0$). In these three series of runs the lane coefficient of friction was held constant at 0.03. Finally, in the fourth series of runs H was measured while the lane coefficient of friction was varied, for constant initial angular velocity, as in the second series of runs, and for $\mathbf{P} = 0$, as in the first series. In all four series, the initial velocity of Q was 6.71 m/sec (22.0 ft/sec) in the $\mathbf{a}_1$ direction, and the initial coordinates of Q were (0., 0.2286, 0.109)m or (0., 0.75, 0.358)ft., see Fig. 5.

The results of the first series of runs are listed in Table 1 where it is seen that the initial angular velocity component most affecting the hook is ω_1. It is interesting to observe that ω_3 which plays a major role in ball-track has little affect on the hook. Table 1 also contains the results of a few runs where ω_2 was independently increased. This results in increased hook verifying the bowling maxim that "staying in the shot (that is, later release and greater initial ω_2) increases the hook."

Since in this case the mass center G and the geometric center Q coincide ($P = 0$), it is possible to obtain a closed-form verification of the entries in Table 1, see reference [10, p. 215]. The entries are indeed found to be consistent with the closed-form solution.

The results of the second series of runs are listed in Table 2 where it is seen that the mass-center positions most affecting hook are generally along the diametral line which is parallel to the initial velocity vector. This result is consistent with intuitive theories of gyrodynamics such as the "law of gyroscopes" or "when forces act upon a spinning body, tending to cause rotation about any other axis than the spinning axis, the spinning axis sets itself in better agreement with the new axis of rotation [11]." It is also consistent with procedures currently used in ball drilling [9].

Similarly, the results of the third series of runs are shown in Table 3 where it is seen that the inertia components most affecting the hook at I_{11} and I_{22}. It is interesting, and also consistent with the results of Table 2, to observe that I_{33} has little effect on the hook.

Table 2   Effect of mass center location on hook

MASS CENTER POSITION (mm)			PERCENT HOOK
P_1	P_2	P_3	H
1.076	-1.076	0.0	0.49
-1.076	1.076	0.0	1.16
0.536	1.426	0.0	1.16
-1.426	0.536	0.0	1.15
-0.878	0.878	0.878	1.22
-0.878	0.878	-0.878	1.00
0.878	-0.878	0.878	0.69
0.878	-0.878	-0.878	0.44
1.524	0.0	0.0	0.62
-1.524	0.0	0.0	1.08
0.0	1.524	0.0	1.10
0.0	-1.524	0.0	0.60
0.0	0.0	-1.524	0.67
1.076	1.076	0.0	0.89
0.0	0.0	1.524	1.08
1.076	1.076	0.0	0.86

Table 3   Effect of inertia components on hook

INERTIA TENSOR I_{ij} (kg m^2)	PERCENT HOOK H
$\begin{bmatrix} 0.0347 & 0.0 & 0.0 \\ 0.0 & 0.0347 & 0.0 \\ 0.0 & 0.0 & 0.0347 \end{bmatrix}$	0.872
$\begin{bmatrix} 0.0347 & -0.0135 & 0.0 \\ -0.0135 & 0.0347 & 0.0 \\ 0.0 & 0.0 & 0.0256 \end{bmatrix}$	0.873
$\begin{bmatrix} 0.0347 & 0.0 & 0.0 \\ 0.0 & 0.0347 & -0.0135 \\ 0.0 & -0.0135 & 0.0347 \end{bmatrix}$	0.790
$\begin{bmatrix} 0.0483 & 0.0 & 0.0 \\ 0.0 & 0.0347 & 0.0 \\ 0.0 & 0.0 & 0.0347 \end{bmatrix}$	0.96
$\begin{bmatrix} 0.0347 & 0.0 & 0.0 \\ 0.0 & 0.0483 & 0.0 \\ 0.0 & 0.0 & 0.0347 \end{bmatrix}$	0.718
$\begin{bmatrix} 0.0347 & 0.0 & 0.0 \\ 0.0 & 0.0347 & 0.0 \\ 0.0 & 0.0 & 0.0483 \end{bmatrix}$	0.872
$\begin{bmatrix} 0.0347 & 0.0 & 0.0 \\ 0.0 & 0.0347 & 0.0 \\ 0.0 & 0.0 & 0.0754 \end{bmatrix}$	0.872
$\begin{bmatrix} 0.0347 & 0.0 & 0.0 \\ 0.0 & 0.0347 & 0.0 \\ 0.0 & 0.0 & 0.1026 \end{bmatrix}$	0.872

The results of the fourth series of runs are shown graphically in Fig. 7, where it is seen that increased lane friction increases the hook. This is also consistent with bowlers' experience [8, 9]. (For the given initial conditions, the curve is valid for $0 \leq \mu < 0.056$. If $\mu \geq 0.056$ rolling occurs before B reaches the pins.)

Conclusions

The foregoing analysis contains several results which might be of interest to analysts, bowling enthusiasts, and professionals: First, the combined use of Euler parameters and angular velocity components as dependent variables is an effective analytical technique for studying the dynamics of complex geometrical systems. Second, of the several factors affecting hook or ball performance, perhaps the most significant is the position of the mass-center with respect to the ball diameter parallel to the initial angular velocity vector as shown in Table 2. This should be encouraging to professionals and others who rely on ball weighting to hopefully improve performance. Finally, the most significant parameter affecting the ball-track is the initial angular velocity component normal to the lane. Interestingly this parameter has relatively little effect on the hook as seen in Table 1.

Acknowledgment

The authors wish to thank Mr. Mark Harlow of the University of Cincinnati Engineering Science Department, in preparing the data for the tables and graph.

References

1 Huston, R. L., and Passerello, C. E., "Eliminating Singularities in Governing Equations of Mechanical Systems," *Mechanics Research Communications*, Vol. 3, No. 5, 1976, pp. 361–365.

2 Whittaker, E. T., *Analytical Dynamics,* Cambridge, London, 1937.

3 Kane, T. R., and Likins, P. W., "Kinematics of Rigid Bodies in Spaceflight," Department of Applied Mechanics, Stanford University, Technical Report No. 204, 1971.

4 Huston, R. L., and Passerello, C. E., "On theDynamics of Chain Systems," ASME Paper No. 74-WA/Aut 11.

5 Passerello, C. E., and Huston, R. L., "An Analysis of General Chain Systems," N72-30532, NASA-CR127924, 1972.

6 Kane, T. R., *Dynamics,* Holt, Rinehart, and Winston, New York, 1968.

7 Eringen, A. C., *Nonlinear Theory of Continuous Media,* McGraw-Hill, New York, 1962.

8 Berger, G., *Bowling for Everyone,* Barnes and Noble, New York, 1973.

9 Johnson, D., *Inside Bowling,* Henry Regency Company, Chicago, Ill., 1973.

10 Kane, T. R., *Analytical Elements of Mechanics,* Vol. 2, Academic Press, New York, 1961.

11 Perry, J., *Spinning Tops and Gyroscopic Motion,* Dover, New York, 1957, p. 26.

APPENDIX

Let the vectors **M** and **W** be defined as

$$\mathbf{M} = M_i \mathbf{a}_i = \boldsymbol{\omega} \times (\mathbf{I} \cdot \boldsymbol{\omega}) \tag{31}$$

and

$$\mathbf{W} = w_i \mathbf{a}_i = \boldsymbol{\omega} \times (\boldsymbol{\omega} \times \mathbf{P}) \tag{32}$$

Then the coefficients a_{ij} and f_i ($i, j = 1, \ldots, 5$) of equations (27) are

$$a_{11} = m; \quad a_{12} = 0; \quad a_{13} = m\mu\tau_1 P_2;$$
$$a_{14} = m(P_3 - \mu\tau_1 P_1); \quad a_{15} = -mP_2$$

$$a_{21} = 0; \quad a_{22} = m; \quad a_{23} = m(-P_3 + \mu\tau_2 P_2);$$
$$a_{24} = -m\mu\tau_2 P_1; \quad a_{25} = mP_1$$

$$a_{31} = 0; \quad a_{32} = -mP_3; \quad a_{33} = I_{11} + m(P_2^2 + P_3^2 + \mu\tau_2 P_2);$$
$$a_{34} = I_{12} - mP_1(P_2 + \mu r\tau_2); \quad a_{35} = I_{13} - mP_1 P_3$$

$$a_{41} = mP_3; \quad a_{42} = 0; \quad a_{43} = I_{21} - mP_2(P_1 + \mu r\tau_1); \quad a_{44}$$
$$= I_{22} + m(P_3^2 + P_1^2 + \mu r\tau_1 P_1); \quad a_{45} = I_{23} - mP_2 P_3$$

$$a_{51} = -mP_2; \quad a_{52} = mP_1; \quad a_{53} = I_{31} - mP_1 P_3; \quad a_{54}$$
$$= I_{32} - mP_2 P_3; \quad a_{55} = I_{33} + m(P_1^2 + P_2^2) \tag{33}$$

and

$$f_1 = -m[w_1 + \mu\tau_1(g + w_3)]; \quad f_2 = -m[w_2 + \mu\tau_2(g + w_3)]; \tag{34}$$

$$f_3 = -M_1 - m[P_2w_3 - P_3w_2 + P_2g + \mu r\tau_2(g + w_3)];$$

$$f_4 = -M_2 - m[P_3w_1 - P_1w_3 - P_1g - \mu r\tau_1(g + w_3)];$$

$$f_5 = -M_3 - m(P_1w_2 - P_2w_1)$$

$$(34)$$

$$(Cont.)$$

where τ_1 and τ_2 are the $\mathbf{a}_1$ and $\mathbf{a}_2$ components of $\boldsymbol{\tau}$ of equation (22).

Similarly, the coefficients b_{ij} and g_j $(i, j = 1, 2, 3)$ of equations (29) and are

$$b_{11} = I_{11} + m[P_2{}^2 + (P_3 + r)^2]; \quad b_{12} = -mP_1P_2;$$

$$b_{13} = -m(P_3 + r)P_1$$

$$(35)$$

$$b_{21} = -mP_1P_2; \quad b_{22} = I_{22} + m[P_1{}^2 + (P_3 + r)^2];$$

$$b_{23} = -m(P_3 + r)P_2$$

$$b_{31} = -mP_1(P_3 + r); \quad b_{32} = -mP_2(P_3 + r);$$

$$b_{33} = I_{33} + m(P_1{}^2 + P_2{}^2)$$

$$(35)$$

$$(Cont.)$$

$$g_1 = -M_1 - m[P_2g + P_2w_3 - (P_3 + r)w_2]$$

$$g_2 = -M_2 + m[P_1g + P_1w_3 - (P_3 + r)w_1]$$

$$g_3 = -M_3 + m[P_2w_1 - P_1w_2]$$

$$(36)$$

Printed in U. S. A.

ERGONOMICS, 1979, VOL. 22, NO. 4, 387–397

Reduction of Wind Resistance and Power Output of Racing Cyclists and Runners Travelling in Groups

By Chester R. Kyle

Mechanical Engineering Department, California State University, Long Beach 90840, U.S.A.

Wind resistance is responsible for most of the metabolic cost of cycling (80–90%), whereas for runners the effect is much less (4–8%). In groups of cyclists and runners, those behind consume less energy, being partially shielded from the wind. Experimental data are presented showing reduction of group wind resistance and power output for cyclists and runners. For cyclists, external power output was reduced over 30% at racing speeds. For runners, energy consumption decreased only 2–4% at middle and long distance speeds. A method of predicting the speed of groups of cyclists and runners is shown. By travelling in a group, cyclists can increase their speed about 0.9–1.8 m s^{-1}, while runners can improve only about 0.1 m s^{-1}. The use of pace lines in cycle racing is by far the most important race tactic. In running, systematic use of pacing is not yet fully utilized although the benefits are potentially significant.

1. Introduction

Over the past several years many investigators have made detailed measurements of the wind and rolling resistance of racing and touring cyclists with the riders using various body positions. From this data they have calculated the external mechanical energy necessary to propel the bicycle at various speeds, (Kyle *et al.* 1973, Kyle 1975, Nonweiler 1956 and 1957, Whitt 1971, Whitt and Wilson 1974). These estimates have correlated fairly well with measurements made on stationary bicycle ergometers, (Harrison 1970, Kyle and Mastropaolo 1976, Pugh 1974, Whitt 1971, Whitt and Wilson 1974).

The case where racing cyclists travel in groups to conserve energy is not as well understood (Kawamura 1953). At speeds over 8.9 m s^{-1}, wind drag comprises over 90% of the total mechanical resistance to motion against a bicycle, (Kyle 1975). Consequently cyclists in the rear of a group may lower their wind resistance significantly by travelling in the wake or 'slipstream', of those in front. By so doing they take advantage of an artificial tailwind; the air is already moving forward when they reach it. Unfortunately there have been few attempts to quantitatively measure the reduction in external power output and changes in energy consumption resulting from this technique called 'slipstreaming' (Kawamura 1953), nor has there been an attempt to model the phenomenon mathematically in order to predict the effect on racing cyclist performance.

The case of runners travelling in groups is somewhat better understood. Pugh (1971) has made a basic effort to measure the effect of one runner shielding another. He reported that at a typical middle distance speed of 6 m s^{-1}, a runner travelling 1 m behind another would consume 6% less total energy. This resulted from an 80% decrease in the metabolic effect of wind resistance (which is about 7.5% of the total energy consumption at this speed). He also estimated that a speed increase of 4 s per 400 m lap might be possible due to shielding. However, this estimate seems to be high compared to an improvement of about 1 s per lap based upon actual race experience, (Pugh 1971). Pugh did not attempt to determine the effect of runner spacing upon the wind resistance. As he noted, it may not be practical for one runner to follow as close as 1 m behind another; and this may have contributed to the over-optimistic estimate of the possible speed increase due to shielding.

The purpose of the present investigation was to measure the effect of 'slipstreaming' upon racing cyclists and runners.

2. Methods

The wind resistance of groups of cyclists and runners could have been measured by putting subjects in a wind tunnel, but this approach proved to be far too complicated and expensive. Some sort of guess might have been made by analysing racing data, but this would only give average values and would not yield enough information. The simplest approach was to use bicycle coasting tests in an enclosed 200 m long hallway employing the methods previously used by Kyle (Kyle *et al.* 1973, Kyle and Edelman 1975). Results from this method compare very well with wind tunnel data, (Kawamura 1953, Nonweiler 1956).

Groups of 1, 2, 3 and 4 cyclists coasted through the instrumented section to measure their rate of deceleration thus permitting the calculation of the total resistance force against each rider in the pace line. In the initial test, resistance forces were measured on cyclists riding in racing position (hands on the lower handle bars, arms stiff, head up). In a second series of tests, the groups coasted through the instrumented section with each rider standing nearly erect in the pedals to approximate the body position of a runner. By subtracting out the known effect of rolling resistance and the wind drag on the bicycle itself, the wind resistance of runners could be estimated (Kyle 1975, Nonweiler 1956).

3. Results

The following factors affecting the wind resistance of groups of cyclists were investigated; the effect of spacing between cyclists, the effect of rider body position, the effect of speed, the effect of place in the pace line and the effect of linear alignment. Others of interest but not investigated were the effect of wind, the effect of closely packed clusters of riders and the effect of differential rider size or varying body position in the same place line.

3.1. *The Effect of Place in the Pace Line*

Results showed that the front cyclist in a pace line was not affected by the process of slipstreaming, although the riders that followed benefitted greatly. If two people are in a pace line at $11 \cdot 1$ m s^{-1} (40 kph), the front person consumes the same energy as if riding alone, while the person following requires about 33 % less power output. At this speed, reduction in wind resistance is about 38 %, however, since rolling resistance is unaffected by slipstreaming, the decrease in external power required is smaller (33 %). Table 1 summarizes these results.

If two slipstreaming cyclists are joined by a third, the power required of the first two remains the same as before. If a fourth person joins the line, the first three experience no change. Also, there seems to be little measurable advantage between positions

Table 1.   Effect of Slipstreaming Upon Bicycles in a Pace Line

Speed kph	24	32	40	48	56
(m s^{-1})	(6·7)	(8·9)	(11·1)	(13·3)	(15·6)
Drop in Wind Resistance	38 %	38 %	38 %	38 %	38 %
Drop in Power Output	29 %	31 %	33 %	34 %	35 %

Riders in racing position with a 0·30 m wheel gap.

Reduced Wind Drag in Groups of Cyclists and Runners 389

in the pace line, as long as a rider is following at least one person. Figure 1 illustrates the drag *vs.* the speed of cyclists in a pace line using racing position. In figure 1, the upper curves represent the total resistance to motion, while the lower curve is the rolling resistance only.

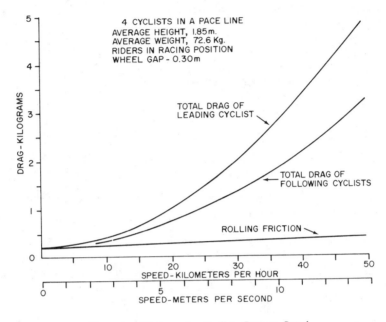

Figure 1. Slipstreaming Cyclists, Drag *vs.* Speed.

3.2. *The Effect of Spacing between Cyclists*

Because of the dimensions of the bicycle, one rider could not approach another closer than about 1·7 m. This represents almost a zero gap between the wheels of the two bicycles. As might be expected, the more closely one cyclist follows another, the greater the drag reduction. Figure 2 summarizes the results of the present investigation and of a wind tunnel study done by Kawamura (1953) using a $\frac{1}{4}$ scale model of two cyclists slipstreaming in racing position. In the present study, the total wind resistance declined an average of 44 % for 1·7 m between riders or a zero wheel gap, and only about 27 % average for 3·7 m between riders or a 2 m wheel gap. Data are not available beyond this point, although obviously a significant drop in wind resistance occurs even with a higher spacing than this.

The curves from the present study represent a second order least squares fit of 99 data points. The wind tunnel experiments by Kawamura (1953), do not agree with the present investigation, showing a higher reduction in wind resistance at all points. At zero wheel gap, for example, for racing position, Kawamura gives 54 % *vs.* a 38 % decrease for the present study. Kawamura's $\frac{1}{4}$ scale models did not have a moving ground plane, which can cause wind tunnel results to vary greatly from free atmosphere tests done with real vehicles. Also, turbulence generated in slipstreaming may not scale upwards linearly as was assumed. Although this may account for some of the difference, part might also be due to imperfect alignment during the coasting tests which was very difficult to control. Regardless of the cause, the discrepancy is not unexpected; Kawamura lists one data point calculated from a field study which is

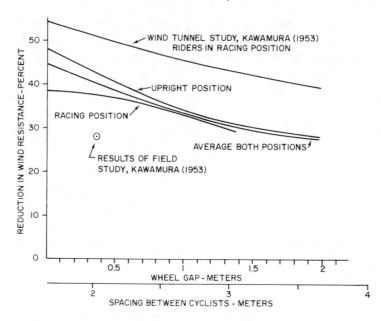

Figure 2. Slipstreaming Cyclists, Reduction in Wind Resistance *vs.* Spacing.

more than 20% below his wind tunnel data (see figure 2). As will be seen later, the present data seem more in accord with field experience.

3.3. *The Effect of Alignment*

If a cyclist overlaps wheels, travelling to one side and behind the rider in front, the benefit from slipstreaming is much less. Although no formal attempt was made to measure this effect, the slipstreaming cyclists unavoidably overlapped during the present tests about 30 times. The decrease in wind resistance varied from zero to 30% (average 23%) depending upon the amount of overlap and the side spacing. This compared to an average of 44% for the case of zero gap with no overlap. Obviously then it is much better to ride directly behind another cyclist using a safe wheel spacing than it is to choose the less advantageous and dangerous overlapping position, (should the front rider swerve suddenly, an accident can easily occur).

It seems reasonable for cyclists riding in the centre of a tightly packed cluster of riders, that wind resistance would be much less than if they were in a straight line as in the present study. However, this effect was not studied in detail. A single test was made with two cyclists in front side by side and one following in a symmetric triangle pattern. Results were surprising; a decrease of only 24% in wind resistance was measured which was about the same as for the case of the overlapping cyclist cited above. Evidently this pattern is not particularly efficient in shielding the rear rider.

3.4. *The Effect of Speed–Drag Coefficient*

The aerodynamic drag coefficient C_d is a measure of the air resistance of a particular geometric shape. Efficient streamlined shapes have low wind resistance and low drag coefficients, often less than 0·1 in magnitude. Bodies that are not streamlined often have drag coefficients 10 times as great as this, sometimes exceeding 1·0. The drag coefficient as used in this paper is defined by the relation:

$$C_d = (2D_w)/(\rho A V^2) \qquad (1)$$

where D_w is the wind drag, ρ is the air density, A is the projected frontal area of the object and V is the relative wind velocity.

Drag coefficients from the present experiments and from Kawamura's (1953) wind tunnel study are shown in table 2. They vary from about 0·8 to about 1·2.

Table 2.   Drag Coefficients *vs.* Velocity, Single Cyclists

Racing Position		Upright Position	
Velocity m s^{-1}	C_d	Velocity m s^{-1}	C_d
4·47 (a)	0·82	4·80 (b)	1·093
5·13 (b)	0·971	6·04 (a)	1·14
6·71 (a)	0·80	7·12 (a)	1·07
7·78 (b)	0·866	7·38 (b)	1·043
8·94 (a)	0·80	8·68 (a)	1·01
11·18 (a)	0·79	9·39 (a)	0·95
11·65 (b)	0·848	11·56 (b)	0·912
20·25 (b)	0·822	20·21 (b)	0·922

(a) Present Study; (b) Kawamura (1953) Wind Tunnel Study.

changing slowly with wind velocity. Pugh (1971) draws a comparison between the wind resistance of a runner and the wind resistance of a cylinder in cross flow. He notes that the drag coefficient of a cylinder is constant for a long period as velocity increases, and then it drops dramatically from about 1·2 to 0·3 as flow in the boundary layer changes from laminar to turbulent. He asks the question whether a similar dramatic drop in wind resistance is experienced by runners as they increase their speed.

In the present experiments the measured drag coefficient for single bicycle riders in the standing position decreased gradually from about 1·14 at 6 m s^{-1} to 0·95 at 9 m s^{-1}. Kawamura's (1953) results are somewhat lower than this (1·09–0·92) for an erect seated posture. With cyclists in racing position, the drag coefficient was nearly constant over the entire range, averaging $C_d = 0·80$. In neither of the positions tested were there any sudden changes in the drag coefficient indicating that rapid transitions in flow characteristics did not occur. From Kawamura's wind tunnel tests it appears that this transition does not exist even at speeds up to 25 m s^{-1}.

Since drag coefficients were nearly constant over the range of speeds tested, it was concluded that average drag coefficients could be used with reasonable accuracy in the case of either single or drafting cyclists or runners regardless of the speed.

3.5. *The Effect of Body Position*

As can be seen from figure 2, the upright body position provided better wind shelter for following riders than did the racing position. This is perfectly reasonable since the upright position is far less streamlined than the racing position, thus generating a greater following air current or 'slipstream'. This should not be interpreted to mean that the upright position has any advantage over the racing position except perhaps comfort. Total drag and energy consumption are still greater under all circumstances in the upright position. It does mean, however, that the larger the body size and frontal area of the leading rider, the greater the advantage to those following in the rear.

All of the observations so far are based on an experimental technique which can measure drag accurate to within about 0·05–0·09 kg. This gives drag measurements accurate to within about ± 3–4% of the total drag. The basic experimental accuracy was adequate to permit the conclusions drawn.

4.   Discussion

Since cyclists at the rear of a pace line consume less energy (see table 1), the group can travel much faster than any single rider if they rotate turns at the front. This is provided all riders are of nearly equal ability. While they are in front or dropping back, the cyclists may go into increasing oxygen debt; however, while in the shelter of the line they are conserving energy reserves for their next turn at the front. From the experimental results it is possible to calculate the increase in speed made possible by slipstreaming.

4.1.   *Predicting the Speed of Cyclists Travelling in Groups*

Let us assume that the speed each rider can maintain alone is known (individual time trial). From this the average external power output of each cyclist P_i may be calculated from this individual's solo performance. This may be found from any one of several experimental studies (Kyle *et al.* 1973, Nonweiler 1957, Whitt 1971, Whitt and Wilson 1974). The long time average power output of all cyclists in a pace line P_a would then be approximately:

$$P_a = \sum_{i=1}^{n} P_i/n \qquad (2)$$

where n is the number of cyclists. If f_h is the fraction of the time riders spend either at the head of the line or dropping back, and f_r is the fraction of the time they spend in the rear shielded from the wind, then the long term average power output of all individuals in the group P_a may also be expressed approximately by:

$$P_a = f_h P_h + f_r P_r \qquad (3)$$

where P_h is the average power output of the cyclist at the head of the line, and P_r is the average power output of each cyclist in the rear. Equations 2 and 3 assume that riders can average the same power output in the group that they can average during an individual time trial under identical conditions. In other words in the pace line their average energy consumption remains about the same as compared to their rate of energy consumption during an individual time trial.

The external power output of a cyclist is approximately a function of the velocity cubed, so that:

$$P_i = CV_i^3 \text{ and } P_h = CV^3 \qquad (4)$$

where V_i is the individual time trial velocity of each rider, V is the velocity of the group, and C is a proportionality constant, (Kyle *et al.* 1973).

From equations 2 and 4 we obtain:

$$P_a = \sum_{i=1}^{n} CV_i^3/n \qquad (5)$$

Equations 3, 4 and 5 then give:

$$\sum_{i=1}^{n} CV_i^3/n = CV^3(f_h + (P_r/P_h)f_r)^{1/3} \qquad (6)$$

and solving for the velocity of the group V results in:

$$V = \left(\sum_{i=1}^{n} V_i^3/n \right)^{1/3} (f_h + (P_r/P_h)f_r)^{-1/3} \tag{7}$$

Now if the riders are of nearly equal ability (their individual time trial speeds are within $\pm 5\%$ of each other), then the root mean cubed is very nearly equal to the geometric average velocity V_a so that:

$$\left(\sum_{i=1}^{n} V_i^3/n \right)^{1/3} \cong \sum_{i=1}^{n} V_i/n = V_a \tag{8}$$

The term $(f_h + (P_r/P_h)f_r)^{-1/3}$ will be called the speed increase factor k. The velocity of the group is then given by the simple expression:

$$V = kV_a \tag{9}$$

If T_a is the simple geometric average of the individual time trial times, then the predicted group time over the same distance is approximately equal to:

$$T = T_a/k \tag{10}$$

This also assumes that the individual time trial times are within $\pm 5\%$ of each other. If this condition is met, then the simplified equations 9 and 10 give answers that are within about $\pm 0{\cdot}1\%$ of those calculated from the more complex equation 7. Also, as will be seen later, it is necessary that riders be of nearly equal ability in order that the pace line speed be greater than that of the best rider. The speed increase factor k is a mild function of velocity and is given in table 3 along with values of f_h, f_r and P_r/P_h. Table 3 assumes a 0·3 m wheel gap (a total spacing of 2 m between riders).

Table 3. Speed Increase Coefficients

Number in Pace Line	f_h	f_r	Speed kph (m s^{-1}) 24 (6·7)	32 (8·9)	40 (11·1)	48 (13·3)	56 (15·6)
			Speed Increase Factor k				
1	1·0	0·0	1·0	1·0	1·0	1·0	1·0
2	0·6	0 4	1 0420	1·0451	1·0483	1·0499	1·0516
3	0·4	0·6	1·0658	1·0710	1·0763	1·0790	1·0817
4	0·3	0·7	1·0786	1·0850	1·0915	1·0948	1·0982
5	0·24	0·76	1·0865	1·0937	1·1010	1·1048	1·1086
P_r/P_h			0·71	0·69	0·67	0·66	0·65

To test the method, data of the USA National and Olympic 4-man 4000 m pursuit teams were used. A summary of the results is shown in table 4. As an example, take the 1975 USA National team. The individuals in the team had 4000 m times of 5:00·84, 5:03·75, 5:10·00 and 5:14·00 (min:s) respectively, for an average time of 5:08·15. This is about 13·0 m s^{-1} or 47·0 kph. Using a speed increase factor of $k = 1{\cdot}0948$ we get a predicted time of 4:40·55 (14·26 m s^{-1} or 51·33 kph) which is one second too low. However, the fourth man dropped out halfway through the race. If we calculate the predicted time for a 3-man team it is 5·1192/1·0790 = 4:44·66. The actual time of 4:41·55 is in between these two times. In the other examples from table 4, errors are both plus and minus, and are quite reasonable considering normal variations in human performance due to motivation and other factors.

Since the model seems to work satisfactorily, we can draw some interesting

394   *Chester R. Kyle*

Table 4.   Computed *vs.* Actual Cycle Race Times

4000 m Pursuit Teams	Average of Individual Times-T_a	Predicted Team Time-T	Actual Race Time	Error s
		Time in Min:s		
1975 U.S.A. Nat. Champions	5:07·15	4:40·55	4:41·55	−1·00
1975 Pan American Games Champions	4:56·10	4:30·46	4:29·10	+1·36
1976 U.S.A. Nat. Champions	5:08·45	4:41·74	4:45·48	−3·74
1976 U.S.A. Nat. Runners Up	5:12·48	4:45·42	4:48·05	−2·63
1976 U.S.A. Olympic Team	5:01·79	4:35·66	4:31·25	+4·41

Speed Increase Factor used $k = 1·0948$.

conclusions about slipstreaming. Take the case of a rider who always follows, thus conserving his energy the entire time. The equations show that they can increase their speed about 1·3–1·8 m s^{-1} (4·8–6·4 kph) over what they are capable of ordinarily. This is provided the pace is steady, on a level course and with little wind. Weaker riders can easily be dropped from the group by any rapid change in pace or condition of the wind or slope.

Next, take the example of pace lines with from two to five members. If all are equally capable of 32 kph (8·9 m s^{-1}) as individual time trialists then their group speeds are as follows:

	1 man	2 man	3 man	4 man	5 man	Lg. Group
Possible Speed kph	32·00	33·44	34·27	34·72	34·99	36·21
(m s^{-1})	(8·89)	(9·29)	(9·52)	(9·64)	(9·72)	(10·06)
Speed Increase kph		1·44	0·83	0·45	0·27	
(m s^{-1})		(0·40)	(0·23)	(0·12)	(0·08)	

Thus, a large group can travel faster than a small group, although the relative speed increase diminishes as the group grows larger.

When riders are not equally matched, the group can actually be slower than the fastest member. Suppose one person can travel 34 kph (9·44 m s^{-1}) while a companion can travel 30 kph (8·33 m s^{-1}); together they will travel only 33·44 kph (9·92 m s^{-1}) if they alternate leads equally. However, in practice the stronger person will usually take a longer lead and thus the group speed can still be higher than any individual in the group.

In the sport of team pursuit, used as an example above, two-four man teams start on the opposite sides of an oval track and race against the clock. They form a precision pace line with very little clearance between wheels, perhaps as little as 15 cm. The equations show that if one team uses 15 cm spacing while another uses 30 cm, the closer spacing will reduce the wind resistance of the following riders in the more closely spaced team by an additional 2 % resulting in a 1·5 to 2 s advantage in 4000 m. This is quite significant in a race. All of the above observations are perfectly in accord with actual race experience. Thus the simple model can provide some interesting quantitative insight into the phenomenon of slipstreaming.

4.2. *Predicting the Speed of Runners Travelling in Groups*

In calculating the speed increase of cyclists, it was possible to use external mechanical power output as a basis, which is quite easily measured due to the unique man–machine combination. With runners this is not possible; measurements of external mechanical power output must be obtained indirectly except in the special case of work done against gravity. In order to derive an equation permitting the calculation of speed increase, it is more convenient to use the total rate of energy consumption E, which may be found from standard metabolic tests. We shall use Margaria's data for the rate of energy consumed during running in still air at zero gradient; a linear algebraic expression for E may be derived using a nomogram in Margaria's paper (Margaria *et al.* 1963):

$$E = 0{\cdot}00775 + 0{\cdot}2485V \qquad (11)$$

where E is in KJ Kg^{-1} min^{-1} and V is in m s^{-1}. In order to estimate the energy consumption due to wind resistance E_w we may use:

$$E_w = (C_d A \rho V^3)/(2FMe) \qquad (12)$$

F is a conversion factor to permit proper choice of units, e is the mechanical efficiency of doing work against wind resistance and M is the mass of the subject. If A is in m^2, M in Kg, ρ in Kg m^{-3} and V in m s^{-1} then $F = 16{\cdot}66$. The average value of $C_d A \rho/2FM$ for all of our subjects was $0{\cdot}0002364 \pm 0{\cdot}00000657$. Pugh (1971) estimates that work done against wind resistance has a mechanical efficiency of $e = 0{\cdot}69$. This gives:

$$E_w = 0{\cdot}0003426V^3 \qquad (13)$$

The total energy consumption with shielding is then approximately:

$$E = 0{\cdot}00775 + 0{\cdot}2485V - 0{\cdot}0003426nV^3 \qquad (14)$$

In equation 14, n is the fractional reduction in wind resistance due to shielding. With no shielding ($n = 0$), at 6 m s^{-1}, a typical middle distance speed, $E = 1{\cdot}499$ KJ Kg^{-1} min^{-1}. We shall assume that the same energy is consumed by the runner with or without shielding. If we use Pugh's (1971) estimate of 80% shielding ($n = 0{\cdot}80$) for one runner following another with a 1 m spacing, then $1{\cdot}499 = 0{\cdot}00775 + 0{\cdot}2485V - 0{\cdot}000274V^3$ from which we find $V = 6{\cdot}27$ m s^{-1}. Pugh (1971) used a different method and calculated a speed of $6{\cdot}4$ m s^{-1}. At a speed of 6 m s^{-1} the time for a 400 m lap without shielding would be $66{\cdot}67$ s and at $6{\cdot}27$ m s^{-1} the time with shielding would be $63{\cdot}79$ s for a saving of $2{\cdot}88$ s per lap. This is somewhat better than Pugh's figure of 4 s per lap, but is still optimistic compared to actual race experience (Pugh 1971).

It should be noted that the energy consumption in equation 13 gives somewhat lower values than those calculated by Pugh (1971), who reports $7{\cdot}5\%$ of the total energy consumption is due to wind resistance at 6 m s^{-1} *vs.* $4{\cdot}9\%$ in the present study. His values were derived from measurements of oxygen consumption while equation 13 resulted from experimentally measured wind resistance. Pugh's value of $e = 0{\cdot}69$ for mechanical efficiency seems unusually high; if a lower value of $e = 0{\cdot}40$ is used, then equation 13 would agree closely with Pugh's work. If a theoretical approach is employed for calculation of the wind drag of runners using the method of Shanebrook (1976) excellent agreement is obtained with the experimental data in this paper. Regardless of the method, the estimate of speed increase using 80% shielding is too high.

From figure 2, a spacing of about 2 m between runners gives a more realistic figure of 40% shielding ($n = 0{\cdot}40$). The speed then becomes $6{\cdot}13$ m s^{-1} for a saving of $1{\cdot}42$ s per lap. This is close to actual race estimates of 1 s per lap (Pugh 1971). In

396   *Chester R. Kyle*

calculating this time saving and in deriving equations 13 and 14, the upright position in cycling was used to provide data on drag coefficients and slipstreaming for runners. This was felt justified since data on single runners agreed so well with cycling data (Shanebrook 1976).

For the mile or metric mile (1500 m), the world record speed is close to 7 m s^{-1}. At this speed 40 % shielding could result in a time advantage of about 1·66 s per lap. This suggests using a national team to break the world mile or 1500 m record. If a record pace could be maintained for 3 plus laps by rotating runners at the front, while shielding the ultimate winning individual, then the winning runner should quite easily be able to improve his best time by several seconds.

At present it is quite common to use special runners to set the pace and help improve race times; however, a more formal and systematic use of pacing could result in substantial improvements on present world marks in middle and even long distance races.

La résistance au vent est principalement responsable de l'accroisement du coût métabolique (80 à 90 %) lors des déplacements en bicyclette, alors que cet effet est nettement moindre chez les coureurs à pieds (4 à 8 %).

Chez les cyclistes et les coureurs qui se déplacent en groupes, ceux qui courent en arrière dépensent moins d'énergie, car ils sont partiellement protégés contre le vent. Cet article fournit des données expérimentales qui démontrent l'importance de cette réduction de la résistance et de la dépense énergétique aussi bien chez des cyclistes que chez des coureurs. Chez les cyclistes, la dépense énergétique est réduite de plus de 30 % pour les vitesses d'une course. Chez les coureurs à pieds, la consommation d'énergie n'est réduite que de 2 à 4 % pour des vitesses de course à moyenne et à longue distance. Une méthode de prédiction de la vitesse dans des groupes de cyclistes et de coureurs a été élaborée. Dans le déplacement en groupes, les cyclistes peuvent accroître leur vitesse d'environ 0,9 à 1,8 m.s^{-1}, alors que les coureurs ne peuvent l'accroître que de 0,1 m.s^{-1} environ. Le recours aux couloirs de course constitue de loin la meilleure tactique pour les courses de bicyclette. Dans les courses à pieds, l'utilisation des couloirs n'est pas systématique, bien que le bénéfice à en retirer soit potentiellement important.

Der Luftwiderstand ist die Hauptursache des Energieverbrauchs beim Radfahren (80–90 %), während seine Auswirkungen bei Läufern sehr viel geringer ist (4–8 %). In Gruppen von Läufern oder Fahrern verbrauchen die hinteren wegen des Windschattens weniger Energie. In der vorliegenden Untersuchung wurden bei Läufern und Fahrern die Verminderung des Luftwiderstandes in der Gruppe und die Leistungsabgabe untersucht. Bei Fahrern wurde die Leistungsabgabe bei Wettkampfgeschwindigkeiten um mehr als 30 % vermindert. Bei Läufern reduzierte sich der Energieverbrauch nur um 2–4 % bei Langstreckengeschwindigkeit. Eine Methode die Geschwindigkeit von Läufer- oder Fahrergruppen vorherzusagen wird vorgestellt. In einer Gruppe kann ein Fahrer seine Geschwindigkeit um 0,9–1,8 m/s erhöhen, für Läufer gilt ein Wert von ca. 0,1 m/s. Beim Radrennen ist die Ausnutzung eines Schrittmachers die bei weitem wichtigste taktische Maßnahme. Bei den Laufwettbewerben wird von der Schrittmacherfunktion trotz ihrer potentiellen Vorteile noch nicht genügend Gebrauch gemacht.

References

HARRISON, J. Y., 1970, Maximizing human power output by suitable selection of motion cycle and load. *Human Factors*, **12**, 315–329.

KAWAMURA, T. M., Nov. 1953, Wind Drag of Bicycles. *Report No. 1, Tokyo University*.

KYLE, C. R., CRAWFORD, C., and NADEAU, D., Nov. 2, 1973, Factors affecting the speed of a bicycle. *Engineering Report 73–1, California State University, Long Beach*.

KYLE, C. R., and EDELMAN, W. E., 1975, Man-powered vehicle design criteria. *Proceedings Third International Conference on Vehicle System Dynamics*. (Amsterdam: SWETS AND ZEITLINGER). Pp. 20–30.

KYLE, C. R., 1975, The aerodynamics of man-powered land vehicles. *Proceedings Seminar on Planning, Design, and Implementation of Bicycle/Pedestrian Facilities*. (Berkeley: INSTITUTE OF TRANSPORTATION AND TRAFFIC ENGINEERING, UNIVERSITY OF CALIFORNIA). Pp. 312–326.

KYLE, C. R., and MASTROPAOLO, J., July 1976, Predicting racing bicyclist performance using the unbraked flywheel method of bicycle ergometry. To be published in the *Proceedings of the International Congress of Physical Activity Sciences, Quebec City, Canada*.

Reduced Wind Drag in Groups of Cyclists and Runners 397

MARGARIA, R., CERRETELLI, P., AGHEMO, P., and SASSI, G., 1963, Energy cost of running. *Journal of Applied Physiology*, **18,** 367–370.

NONWEILER, T., Oct. 1956, The air resistance of racing cyclists. *Report No. 106, The College of Aeronautics, Cranfield, England.*

NONWEILER, T., May 10, 1957, Power output of racing cyclists. *Engineering*, **181,** 586.

PUGH, L. G. C. E., 1971, The influence of wind resistance in running and walking and the mechanical efficiency of work against horizontal or vertical forces. *Journal of Physiology*, (*London*) **213,** 255–276.

PUGH, L. G. C. E., 1974, The relation of oxygen intake and speed in competiton cycling and comparative observations on the bicycle ergometer. *Journal of Physiology*, (*London*) **241,** 795–808.

SHANEBROOK, J. R., and JASZCZAK, R. D., 1976, Aerodynamic drag analysis of runners. *Medicine and Science in Sports*, **8,** 43–45.

WHITT, F. R., 1971, A note on the estimation of the energy expenditure of sporting cyclists. *Ergonomics*, **14,** 419–424.

WHITT, F. R. and WILSON, D. G., 1974, *Bicycling Science, Ergonomics and Mechanics*. (Cambridge: MASSACHUSETTS INSTITUTE OF TECHNOLOGY PRESS).

Manuscript received 31 January 1978.
Revised Manuscript received 3 March 1978.

Reprinted by permission from *Nature*, Vol. **85**, No., 2147, pp. 2151-2157. ©1910 Macmillan Journals Limited.

THE DYNAMICS OF A GOLF BALL.[1]

THERE are so many dynamical problems connected with golf that a discussion of the whole of them would occupy far more time than is at my disposal this evening. I shall not attempt to deal with the many important questions which arise when we consider the impact of the club with the ball, but confine myself to the consideration of the flight of the ball after it has left the club. This problem is in any case a very interesting one; it would be even more interesting if we could accept the explanations of the behaviour of the ball given by many contributors to the very voluminous literature which has collected round the game; if these were correct, I should have to bring before you this evening a new dynamics, and announce that matter, when made up into golf balls, obeys laws of an entirely different character from those governing its action when in any other condition.

If we could send off the ball from the club, as we might from a catapult, without spin, its behaviour would be regular, but uninteresting; in the absence of wind its path would keep in a vertical plane; it would not deviate

FIG. 1.

either to the right or to the left, and would fall to the ground after a comparatively short carry.

But a golf ball when it leaves the club is only in rare cases devoid of spin, and it is spin which gives the interest, variety, and vivacity to the flight of the ball. It is spin which accounts for the behaviour of a sliced or pulled ball, it is spin which makes the ball soar or "douk," or execute those wild flourishes which give the impression that the ball is endowed with an artistic temperament, and performs these eccentricities as an acrobat might throw in an extra somersault or two for the fun of the thing. This view, however, gives an entirely wrong impression of the temperament of a golf ball, which is, in reality, the most prosaic of things, knowing while in the air only one rule of conduct, which it obeys with unintelligent conscientiousness, that of always following its nose. This rule is the key to the behaviour of all balls when in the air, whether they are golf balls, base balls, cricket balls, or tennis balls. Let us, before entering into the reason for this rule, trace out some of its consequences. By the nose of the ball we mean the point on the ball furthest in front. Thus if, as in Fig. 1, C the centre of the ball is moving horizontally to the right, A will be the nose of the ball; if it is moving horizontally to the left, B will

[1] Discourse delivered at the Royal Institution on Friday. March 18, by Sir J. J. Thomson, F.R.S.

252　　　　　　　　　*NATURE*　　　　　　[DECEMBER 22, 1910

be the nose. If it is moving in an inclined direction CP, as in Fig. 2, then A will be the nose.

Now let the ball have a spin on it about a horizontal axis, and suppose the ball is travelling horizontally as in Fig. 3, and that the direction of the spin is as in the

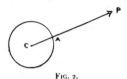

FIG. 2.

figure, then the nose A of the ball is moving upwards, and since by our rule the ball tries to follow its nose, the ball will rise and the path of the ball will be curved as in the dotted line. If the spin on the ball, still about a horizontal axis, were in the opposite direction, as in

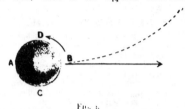

FIG. 3.

Fig. 4, then the nose A of the ball would be moving downwards, and as the ball tries to follow its nose it will duck downwards, and its path will be like the dotted line in Fig. 4.

Let us now suppose that the ball is spinning about a

FIG. 4.

vertical axis, then if the spin is as in Fig. 5, as we look along the direction of the flight of the ball the nose is moving to the right; hence by our rule the ball will move off to the right, and its path will resemble the dotted line in Fig. 5; in fact, the ball will behave like a sliced ball.

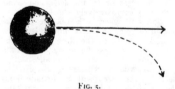

FIG. 5.

Such a ball, as a matter of fact, has spin of this kind about a vertical axis.

If the ball spins about a vertical axis in the opposite direction, as in Fig. 6, then, looking along the line of flight, the nose is moving to the left, hence the ball moves

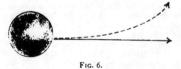

FIG. 6.

off to the left, describing the path indicated by the dotted line; this is the spin possessed by a " pulled " ball.

If the ball were spinning about an axis along the line of flight, the axis of spin would pass through the nose of the ball, and the spin would not affect the motion of

the nose; the ball, following its nose, would thus move on without deviation.

Thus, if a cricket ball were spinning about an axis parallel to the line joining the wickets, it would not swerve in the air; it would, however, break in one way or the other after striking the ground; if, on the other hand, the ball were spinning about a vertical axis, it would swerve while in the air, but would not break on hitting the ground. If the ball were spinning about an axis intermediate between these directions it would both swerve and break.

Excellent examples of the effect of spin on the flight of a ball in the air are afforded in the game of base ball; an expert pitcher, by putting on the appropriate spins, can make the ball curve either to the right or to the left, upwards or downwards; for the sideway curves the spin must be about a vertical axis, for the upward or downward ones about a horizontal axis.

A lawn-tennis player avails himself of the effect of spin when he puts " top spin " on his drives, i.e. hits the ball on the top so as to make it spin about a horizontal axis, the nose of the ball travelling downwards, as in Fig. 4; this makes the ball fall more quickly than it otherwise would, and thus tends to prevent it going out of the court.

Before proceeding to the explanation of this effect of spin, I will show some experiments which illustrate the point we are considering. As the forces acting on the ball depend on the *relative* motion of the ball and the air, they will not be altered by superposing the same velocity on the air and the ball; thus, suppose the ball is rushing forward through the air with the velocity V, the forces will be the same if we superpose on both air and ball a velocity equal and opposite to that of the ball; the effect of this is to reduce the centre of the ball to rest, but to

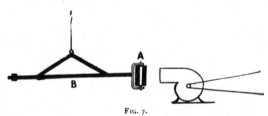

FIG. 7.

make the air rush past the ball as a wind moving with the velocity V. Thus the forces are the same when the ball is moving and the air at rest, or when the ball is at rest and the air moving. In lecture experiments it is not convenient to have the ball flying about the room; it is much more convenient to keep the ball still and make the air move.

The first experiment I shall try is one made by Magnus in 1852; its object is to show that a rotating body moving relatively to the air is acted on by a force in the direction in which the nose of the body is moving relatively to its centre; the direction of this force is thus at right angles both to the direction in which the centre of the body is moving and also to the axis about which the body is spinning. For this purpose a cylinder A (Fig. 7) is mounted on bearings so that it can be spun rapidly about a vertical axis; the cylinder is attached to one end of the beam B, which is weighted at the other end, so that when the beam is suspended by a wire it takes up a horizontal position. The beam yields readily to any horizontal force, so that if the cylinder is acted on by such a force this will be indicated by the motion of the beam. In front of the cylinder there is a pipe D, through which a rotating fan driven by an electric motor sends a blast of air which can be directed against the cylinder. I adjust the beam and the beam carrying the cylinder so that the blast of air strikes the cylinder symmetrically; in this case, when the cylinder is not rotating the impact against it of the stream of air does not give rise to any motion of the beam. I now spin the cylinder, and you see that when the blast strikes against it the beam moves off sideways. It goes off one way when the spin is in one direction, and in the opposite way when the direction of spin is reversed.

DECEMBER 22, 1910]   *NATURE*   253

The beam, as you will see, rotates in the same direction as the cylinder, which an inspection of Fig 8 will show you is just what it would do if the cylinder were acted upon by a force in the direction in which its nose (which, in this case, is the point on the cylinder first struck by the blast) is moving. If I stop the blast the beam does not move, even though I spin the cylinder, nor does it move when the blast is in action if the rotation of the cylinder is stopped; thus both spin of the cylinder and

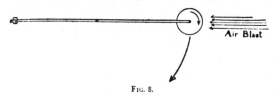

FIG. 8.

movement of it through the air are required to develop the force on the cylinder.

Another way of showing the existence of this force is to take a pendulum the bob of which is a cylinder, or some other symmetrical body, mounted so that it can be set in rapid rotation about a vertical axis. When the bob of the pendulum is not spinning the pendulum keeps swinging in one plane, but when the bob is set spinning the plane in which the pendulum swings no longer remains stationary, but rotates slowly in the same sense as the bob is spinning (Fig. 9).

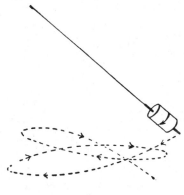

F G. 9.

We shall now pass on to the consideration of how these forces arise. They arise because when a rotating body is moving through the air the pressure of the air on one side of the body is not the same as that on the other; the pressures on the two sides do not balance, and thus the body is pushed away from the side where the pressure is greatest.

Thus, when a golf ball is moving through the air, spinning in the direction shown in Fig. 10, the pressure

FIG. 10.

on the side ABC, where the velocity due to the spin conspires with that of translation, is greater than that on the side ADB, where the velocity due to the spin is in the opposite direction to that due to the translatory motion of the ball through the air.

I will now try to show you an experiment which proves that this is the case, and also that the difference between the pressure on the two sides of the golf ball depends upon the roughness of the ball.

In this instrument, Fig. 11, two golf balls, one smooth

and the other having the ordinary bramble markings, are mounted on an axis, and can be set in rapid rotation by an electric motor. An air-blast produced by a fan comes through the pipe B, and can be directed against the balls; the instrument is provided with an arrangement by which the supports of the axis carrying the balls can be raised or lowered so as to bring either the smooth or the bramble-marked ball opposite to the blast. The pressure is measured in the following way:—LM are two tubes connected with the pressure-gauge PQ; L and M are placed so that the golf balls can just fit in between them; if the pressure of the air on the side M of the balls is greater than that of the side L, the liquid on the right-hand side Q of the pressure-gauge will be depressed; if, on the other hand, the pressure at L is greater than that at M, the left-hand side P of the gauge will be depressed.

I first show that when the golf balls are not rotating there is no difference in the pressure on the two sides when the blast is directed against the balls; you see there is no motion of the liquid in the gauge. Next I stop the blast and make the golf balls rotate; again there is no motion in the gauge. Now when the golf balls are spinning in the direction indicated in Fig. 11 I turn on the blast, the liquid falls on the side Q of the gauge, rises on the other side. Now I reverse the direction of rotation of the balls, and you see the motion of the liquid in the gauge is reversed, indicating that the high pressure has gone from one side to the other. You see that the pressure is higher on the side M, where the spin carries this side of the ball into the blast, than on L, where the spin tends to carry the ball away from the blast. If we could

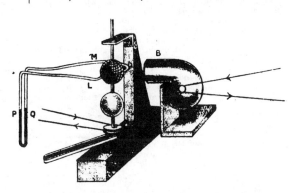

FIG. 11.

imagine ourselves on the golf ball, the wind would be stronger on the side M than on L, and it is on the side of the strong wind that the pressure is greatest. The case when the ball is still and the air moving from right to left is the same from the dynamical point of view as when the air is still and the ball moves from left to right; hence we see that the pressure is greatest on the side where the spin makes the velocity through the air greater than it would be without spin.

Thus, if the golf ball is moving as in Fig. 12, the spin increases the pressure on the right of the ball and diminishes the pressure on the left.

To show the difference between the smooth ball and the rough one, I bring the smooth ball opposite the blast; you observe the difference between the levels of the liquid in the two arms of the gauge. I now move the rough ball into the place previously occupied by the smooth one, and you see that the difference of the levels is more than doubled, showing that with the same spin and speed of air blast the difference of pressure for the rough ball is more than twice that for the smooth.

We must now go on to consider why the pressure of the air on the two sides of the rotating ball should be different. The gist of the explanation was given by Newton nearly 250 years ago. Writing to Oldenburg in 1671 about the dispersion of light, he says, in the course of his letter:—
" I remembered that I had often seen a tennis ball struck with an oblique racket describe such a curved line. For

²54   *NATURE*   [December 22, 1910

a circular as well as progressive motion being communicated to it by that stroke, its parts on that side where the motions conspire must press and beat the contiguous air more violently, and there excite a reluctancy and reaction of the air proportionally greater." This letter has more than a scientific interest—it shows that Newton set an excellent precedent to succeeding mathematicians and physicists by taking an interest in games. The same explanation was given by Magnus, and the mathematical theory of the effect is given by Lord Rayleigh in his paper on "The Irregular Flight of a Tennis Ball," published in the *Messenger of Mathematics*, vol. vi., p. 14, 1877. Lord Rayleigh shows that the force on the ball resulting from this pressure difference is at right angles to the direction of motion of the ball, and also to the axis of spin, and that the magnitude of the force is proportional to the velocity of the ball multiplied by the velocity of spin, multiplied by the sine of the angle between the direction of motion of the ball and the axis of spin. The analytical investigation of the effects which a force of this type would produce on the movement of a golf ball has been discussed very fully by Prof. Tait, who also made a very interesting series of experiments on the velocities and spin of golf balls when driven from the tee, and the resistance they experience when moving through the air.

Fig. 12.

As I am afraid I cannot assume that all my hearers are expert mathematicians, I must endeavour to give a general explanation, without using symbols, of how this difference of pressure is established.

L t us consider a golf ball (Fig. 13) rotating in a current of air flowing past it. The air on the lower side of the ball will have its motion checked by the rotation of the ball, and will thus in the neighbourhood of the ball move more slowly than it would do if there were no golf ball present, or than it would do if the golf ball were there but was not spinning. Thus if we consider a stream of air flowing along the channel PQ, its velocity when near the ball at Q must be less than its velocity when it started at P; there must, then, have been pressure acting against the motion of the air as it moved from P to Q, i.e. the pressure of the air at Q must be greater than at a place like P, which is some distance from the ball. Now let us consider the other side of the ball; here the spin tends to carry the ball in the direction of the blast of air; if the velocity of the surface of the ball is greater than that of the blast, the ball will increase the velocity of the blast on this side, and if the velocity of the ball is less than that of the blast, though it will diminish the velocity of the air, it will not do so to so great an extent as on the other side of the ball. Thus the increase in pressure of the air at the top of the ball over that at P, if it exists at all, will be less than the increase in pressure at the bottom of the ball. Thus the pressure at the bottom of the ball will be greater than that at the top, so that the ball will be acted on by a force tending to make it move upwards.

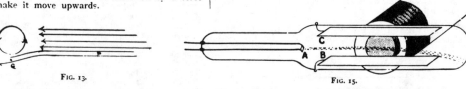

Fig. 13.

We have supposed here that the golf ball is at rest, and the air rushing past it from right to left; the forces are just the same as if the air were at rest, and the golf ball rushing through it from left to right. As in Fig. 13, such a ball rotating in the direction shown in the figure will move upwards, i.e. it will follow its nose.

It may perhaps make the explanation of this difference of pressure easier if we take a somewhat commonplace example of a similar effect. Instead of a golf ball, let us consider the case of an Atlantic liner, and, to imitate the rotation of the ball, let us suppose that the passengers are taking their morning walk on the promenade deck, all circulating round the same way. When they are on one

side of the boat they have to face the wind, on the other side they have the wind at their backs. Now when they face the wind, the pressure of the wind against them is greater than if they were at rest, and this increased pressure is exerted in all directions, and so acts against the part of the ship adjacent to the deck; when they are moving with their backs to the wind the pressure against their backs is not so great as when they were still, so the pressure acting against this side of the ship will not be so great. Thus the rotation of the passengers will increase the pressure on the side of the ship when they are facing the wind and diminish it on the other side. This case is quite analogous to that of the golf ball.

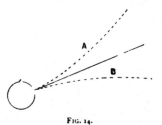

Fig. 14.

The difference between the pressures on the two sides of the golf ball is proportional to the velocity of the ball multiplied by the velocity of the spin. As the spin imparted to the ball by a club with a given loft is proportional to the velocity with which the ball leaves the club, the difference of pressure when the ball starts is proportional to the square of its initial velocity. The difference between the average pressures on the two sides of the ball need only be about one-fifth of 1 per cent. of the atmospheric pressure to produce a force on the ball greater than its weight. The ball leaves the club in a good drive with a velocity sufficient to produce far greater pressures than this. The consequence is that when the ball starts from the tee spinning in the direction shown in Fig. 14, this is often called underspin; the upward force due to the spin is greater than its weight, thus the resultant force is upwards, and the ball is repelled from the earth instead of being attracted to it. The consequence is that the path of the ball curves upward as in the curve A instead of downwards as in B, which would be its path if it had no spin. The spinning golf ball is, in fact, a very efficient heavier-than-air flying machine; the lifting force may be many times the weight of the ball.

The path of the golf ball takes very many interesting forms as the amount of spin changes. We can trace all these changes in the arrangement which I have here, and which I might call an electric golf links. With this apparatus I can subject small particles to forces of exactly the same type as those which act on a spinning golf ball.

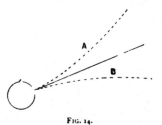

Fig. 15.

These particles start from what may be called the tee A (Fig. 15). This is a red-hot piece of platinum with a spot of barium oxide upon it; the platinum is connected with an electric battery which causes negatively electrified particles to fly off the barium and travel down the glass tube in which the platinum strip is contained; nearly all the air has been exhausted from this tube. These particles are luminous, so that the path they take is very easily observed. We have now got our golf balls off from the tee; we must now introduce a vertical force to act upon them to correspond to the force of gravity on the golf ball. This is easily done by the horizontal plates BC, which are electrified by connecting them with an electric

DECEMBER 22, 1910]   *NATURE*   255

battery; the upper one is electrified negatively, hence when one of these particles moves between the plates it is exposed to a constant downwards force, quite analogous to the weight of the ball. You see now when the particles pass between the plates their path has the shape shown in Fig. 16; this is the path of a ball without spin. I can imitate the effect of spin by exposing the particles while they are moving to magnetic force, for the theory of these particles shows that when a magnetic force acts upon them it produces a mechanical force which is at right angles

FIG. 16.

to the direction of motion of the particles, at right angles also to the magnetic force, and proportional to the product of the velocity of the particles, the magnetic force, and the sine of the angle between them. We have seen that the force acting on the golf ball is at right angles to the direction in which it is moving at right angles to the axis of spin, and proportional to the product of the velocity of the ball, the velocity of spin, and the sine of

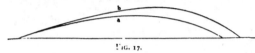

FIG. 17.

the angle between the velocity and the axis of spin. Comparing these statements, you will see that the force on the particle is of the same type as that on the golf ball if the direction of the magnetic force is along the axis of spin and the magnitude of the force proportional to the velocity of spin, and thus if we watch the behaviour of these particles when under the magnetic force we shall get an indication of the behaviour of the spinning golf

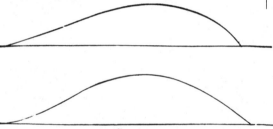

FIG. 18.

ball. Let us first consider the effect of underspin on the flight of the ball; in this case the ball is spinning, as in Fig. 3, about a horizontal axis at right angles to the direction of flight. To imitate this spin I must apply a horizontal magnetic force at right angles to the direction

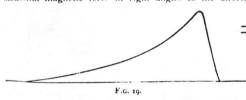

FIG. 19.

of flight of the particles. I can do this by means of the electromagnet. I will begin with a weak magnetic force, representing a small spin. You see how the path differs from the one when there was no magnetic force; the path, to begin with, is flatter, though still concave, and the carry is greater than before—see Fig. 17, *a*. I now increase the strength of the magnetic field, and you will see that the carry is still further increased, Fig. 17, *b*. I increase the spin still further, and the initial path becomes convex instead of concave, with a still further increase in carry, Fig. 18. Increasing the force still

more, you see the particle soars to a great height, then comes suddenly down, the carry now being less than in the previous case (Fig. 19). This is still a familiar type of the path of the golf ball. I now increase the magnetic force still further, and now we get a type of flight not to my knowledge ever observed in a golf ball, but which would be produced if we could put on more spin than

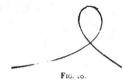

FIG. 20.

we are able to do at present. You see there is a kink in the curve, and at one part of the path the particle is actually travelling backwards (Fig. 20). Increasing the magnetic force I get more kinks, and we have a type of drive which we have to leave to future generations of golfers to realise (Fig 21).

FIG. 21.

By increasing the strength of the magnetic field I can make the curvature so great that the particles fly back behind the tee, as in Fig. 22.

So far I have been considering underspin. Let us now illustrate slicing and pulling; in these cases the ball is spinning about a vertical axis. I must therefore move my electromagnet, and place it so that it produces a vertical magnetic force (Fig. 23). I make the force act

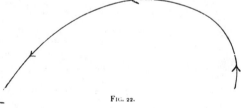

FIG. 22.

one way, say downwards, and you see the particles curve away to the right, behaving like a sliced ball. I reverse the direction of the force and make it act upwards, and the particles curve away to the left, just like a pulled ball.

By increasing the magnetic force we can get slices and

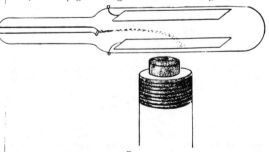

FIG. 23.

pulls much more exuberant than even the worst we perpetrate on the links.

Though the kinks shown in Fig. 20 have never, so far as I am aware, been observed on a golf links, it is quite easy to produce them if we use very light balls. I have

256   *NATURE*   [DECEMBER 22, 1910

here a ball A made of very thin indiarubber of the kind used for toy balloons, filled with air, and weighing very little more than the air it displaces; on striking this with the hand, so as to put underspin upon it, you see that it describes a loop, as in Fig. 24.

Striking the ball so as to make it spin about a vertical axis, you see that it moves off with a most exaggerated slice when its nose is moving to the right looking at it from the tee, and with an equally pronounced pull when its nose is moving to the left.

One very familiar property of slicing and pulling is that the curvature due to them becomes much more pronounced when the velocity of the ball has been reduced than it was at the beginning when the velocity was greatest. We can easily understand why this should be so if we consider the effect on the sideways motion of reducing the velocity to one half. Suppose a ball is pro-

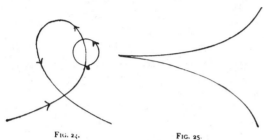

FIG. 24.   FIG. 25.

jected from A in the direction AB, but is sliced; let us find the sideways motion BC due to slice. The sideways force is, as we have seen, proportional to the product of the velocity of the ball and the velocity of spin, or, if we keep the spin the same in the two cases, to the velocity of the ball; hence, if we halve the velocity we halve the sideways force, hence, in the same time, the displacement would be halved too, but when the velocity is halved the time taken for the ball to pass from A to B is doubled. Now the displacement produced by a constant force is proportional to the square of the time; hence, if the force had remained constant, the sideways deflection BC would have been increased four times by halving the velocity, but as halving the velocity halves the force, BC is doubled when the velocity is halved; thus the sideways movement is twice as great when the velocity is halved.

If the velocity of the spin diminished as rapidly as that of translation, the curvature would not increase as the velocity diminished, but the resistance of the air has more effect on the speed of the ball than on its spin, so that the speed falls the more rapidly of the two.

The general effect of wind upon the motion of a spinning ball can easily be deduced from the principles we discussed in the earlier part of the lecture. Take, first, the case of a head-wind. This wind increases the relative velocity of the ball with respect to the air; since the force due to the spin is proportional to this velocity, the wind

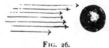

FIG. 26.

increases this force, so that the effects due to spin are more pronounced when there is a head-wind than on a calm day. All golfers must have had only too many opportunities of noticing this. Another illustration is found in cricket; many bowlers are able to swerve when bowling against the wind who cannot do so to any considerable extent on a calm day.

Let us now consider the effect of a cross-wind. Suppose the wind is blowing from left to right, then, if the ball is pulled, it will be rotating in the direction shown in Fig. 26; the rules we found for the effect of rotation on the difference of pressure on the two sides of a ball in a blast of air show that in this case the pressure on the front half of the ball will be greater than that on the rear half, and thus tend to stop the flight of the ball. If,

however, the spin was that for a slice, the pressure on the rear half would be greater than the pressure in front, so that the difference in pressure would tend to push on the ball and make it travel further than it otherwise would. The moral of this is that if the wind is coming from the left we should play up into the wind and slice the ball, while if it is coming from the right we should play up into it and pull the ball.

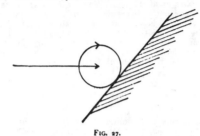

FIG. 27.

I have not time for more than a few words as to how the ball acquires the spin from the club. But if you grasp the principle that the action between the club and the ball depends only on their *relative* motion, and that it is the same whether we have the ball fixed and move the club or have the club fixed and project the ball against it, the main features are very easily understood.

Suppose Fig. 27 represents the section of the head of a

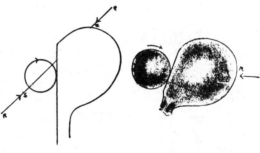

FIG. 28.   FIG. 29.

lofted club moving horizontally forward from right to left, the effect of the impact will be the same as if the club were at rest and the ball were shot against it horizontally from left to right. Evidently, however, in this case the ball would tend to roll up the face, and would thus get spin about a horizontal axis in the direction shown in the figure; this is underspin, and produces the upward force which tends to increase the carry of the ball.

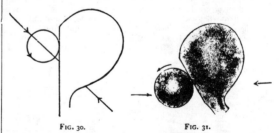

FIG. 30.   FIG. 31.

Suppose, now, the face of the club is not square to its direction of motion, but that, looking down on the club, its line of motion when it strikes the ball is along PQ (Fig. 28), such a motion as would be produced if the arms were pulled in at the end of the stroke, the effect of the impact now will be the same as if the club were at rest and the ball projected along RS, the ball will endeavour to roll along the face away from the striker; it will spin

in the direction shown in the figure about a vertical axis. This, as we have seen, is the spin which produces a slice. The same spin would be produced if the motion of the club was along LM and the face turned so as to be in the position shown in Fig. 29, *i.e.* with the heel in front of the toe.

If the motion and position of the club were as in Figs. 30 and 31, instead of as in Figs. 28 and 29, the same consideration would show that the spin would be that possessed by a pulled ball.

Golf Ball Aerodynamics

P W BEARMAN AND J K HARVEY

(Imperial College of Science and Technology)

Summary: A wind tunnel technique has been developed to measure the aerodynamic forces acting on golf balls over a wide range of Reynolds number and spin rate. Balls with round dimples and hexagonal dimples have been investigated. The dimples are found to induce a critical Reynolds number behaviour at a lower value of Reynolds number than that experienced by a smooth sphere and beyond this point, unlike the behaviour of a sand-roughened sphere, there is little dependence of the forces on further increases in Reynolds number. A hexagonally-dimpled ball has a higher lift coefficient and a slightly lower drag coefficient than a conventional round-dimpled ball. Trajectories are calculated using the aerodynamic data and the ranges are compared with data obtained from a driving machine on a golf course.

1.   Introduction

Golf, in common with many other games, is strongly influenced by the aerodynamic forces that act on a sphere. After the introduction of the gutta-percha ball in 1845, golfers discovered that it flew farther and better when scored or marked. Thus started the introduction of numerous cover designs chosen more or less by intuition. Among those tried included the "bramble ball" with a raised pattern, which unfortunately tended to accumulate mud, and covers with rectangular and square depressions. By 1930 the round dimple had almost completely taken over and became accepted as the standard design for golf balls. These "conventional" golf balls have either 330 or 336 round dimples placed in regular rows.

The trajectory of a golf ball is determined by the gravitational and aerodynamic forces acting on it during its flight. In addition to the drag, a lift force is generated by the back spin imparted to the ball by the angled face of the club. Initial rotational speeds of between 2000 and 4000 rpm have been measured for typical drive shots. The effect of the spin is to delay separation of the flow from the upper part of the surface and to advance it on the lower. The aerodynamic lift L and drag D acting on a spinning sphere depend on the velocity through the air U, the fluid density ρ and the viscosity μ. In addition they will depend on the diameter d and the rotational speed N, measured in revolutions per minute. It has been found convenient to express the spin in terms of the peripheral or equatorial speed v, where $v = \pi Nd/60$. Therefore we can write

$$L \text{ and } D = f(\rho, \mu, U, d, v).$$

From the rules of dimensional analysis, this equation becomes

$$C_L \text{ and } C_D = f\left(\frac{\rho Ud}{\mu}, \frac{v}{U}\right),$$

where $\rho Ud/\mu$ is the Reynolds number Re and the velocity ratio v/U will be termed the spin parameter.

$$C_L = \frac{L}{\frac{1}{2}\rho U^2 S} \text{ and } C_D = \frac{D}{\frac{1}{2}\rho U^2 S}$$

where $S = \pi d^2/4$, the projected area of the sphere. During a typical drive both the Reynolds number and the spin parameter will change and it is expected that the lift and drag coefficients will also vary. The spin parameter will be about 0.1 if the initial rpm of the ball is around 3500. During the flight the velocity of the ball will fall, thus reducing the Reynolds number, and, although there will be some decay of the spin, the drop in velocity will cause the spin parameter to increase.

Received January 1976

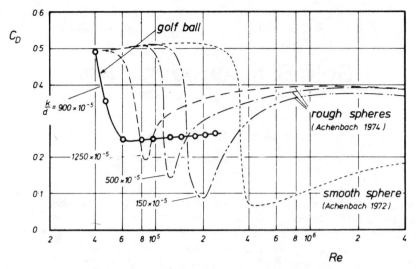

Figure 1   Variation of golf ball and sphere drag

The British golf ball has a diameter of 1.62 in (41.1 mm) and, assuming a speed of 75 m/s off the tee, the initial Reynolds number based on ball diameter is about 2.1×10^5. Figure 1, taken from Achenbach[1], clearly shows that a smooth sphere at this Reynolds number is in the high drag sub-critical flow regime where the boundary layer remains laminar up to separation. The addition of surface roughness, in the form of spherical grains, reduces the critical Reynolds number and, apart from the sphere with finest roughness, the drag coefficient quickly recovers, with increasing Reynolds number, to a value of about 0.4. This helps to explain why a dimpled golf ball can be driven further than a smooth one, but dimples are a very gross roughness and it is doubtful whether they can be compared directly with sand-roughened spheres.

Data on the aerodynamic forces experienced by spheres rotating about an axis normal to the flight direction are sparse. Figure 2 shows wind-tunnel measurements of Maccoll[2] and Davies[3]. C_D and C_L are plotted against the spin parameter v/U for a Reynolds number of about 10^5. Cases 1 and 2 are for a smooth sphere, whereas case 3 gives forces for a conventionally-dimpled golf ball. The accuracy of both sets of data is open to question. Maccoll supported his sphere on a spindle which had a diameter equal to about 15 per cent of the sphere diameter. Davies used a device which spun the ball up between a pair of rotating cups. The cups were parted, releasing the ball into the air stream. By noting the distance travelled by the sphere before it struck the floor of the tunnel, the aerodynamic forces could be deduced. A limitation of Davies's work was that the Reynolds number of his golf ball measurements was 9.4×10^4, i e somewhat less than the value expected of a golf ball leaving the tee. Some values of drag coefficient are presented by Williams[4], but these were deduced from range data assuming that both lift and spin are unimportant.

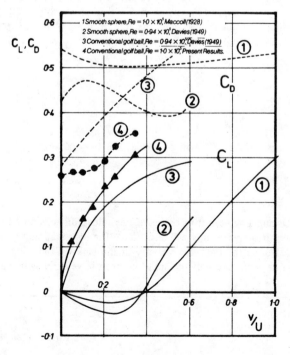

Figure 2   Lift and drag coefficients of rotating spheres versus rotational speed

The main aim of the present investigation was to measure the lift and drag forces acting on a golf ball in a wind tunnel and to use these data to predict golf ball trajectories. The ranges from these predicted trajectories can then be compared with actual ranges measured under controlled conditions. A further aim was to examine the effect on the aerodynamic forces, and hence the range, of changing the dimple shape. The forces on a conventionally-dimpled golf ball were compared with those on a ball having a dimple configuration similar to that on a Uniroyal Plus 6. The Plus 6 has 240 hexagonal dimples and 12 pentagonal dimples arranged in a triangulate pattern. The manufacturers claim that in most circumstances the Plus 6, hereafter called the hexagonally-dimpled ball, will carry farther than a conventional ball when given the same initial speed, spin and launch angle.

Notation

L	lift
D	drag
C_L	lift coefficient
C_D	drag coefficient
ρ	fluid density
μ	fluid viscosity
N	rotational speed
U	velocity through the air
d	diameter
v	peripheral velocity (spin), $=\pi Nd/60$
Re	Reynolds number, $=\rho Ud/\mu$
S	projected area of ball, $=\pi d^2/4$
k	average diameter of sand grains

2. Wind-Tunnel Experiments

2.1 EXPERIMENTAL ARRANGEMENT

The wind-tunnel measurements were made in the Imperial College, Department of Aeronautics 5 ft x 4 ft (1.524 m x 1.219 m) wind tunnel. This is of the closed return type and has a turbulence level of about 0.2 per cent. The maximum speed of the tunnel is around 45 m/s and therefore, in order to simulate the correct Reynolds number, it was necessary to use models of golf balls 2½ times full scale. The tunnel was operated up to speeds of about 37 m/s. This simulates flow on a golf ball at a speed of 92.5 m/s.

Three wind-tunnel models were made: a conventionally-dimpled ball, a hexagonally-dimpled ball and a smooth sphere. Various authors have pointed out the difficulty of accurately measuring the forces on spheres, owing to support interference. Since reliable data exist on the drag of a non-rotating smooth sphere, the results for our smooth sphere could be used to assess the degree of interference caused by the support system. The models were constructed as hollow shells moulded in glass reinforced plastic. Each shell was split in two to accommodate a motor and bearing assembly on which the ball revolved. These details are shown in Figure 3. Various support systems were tried and the one finally adopted was to suspend the ball from a wire of 0.50 mm diameter. A second wire of 0.20 mm diameter was attached to the underside of the ball, passing through the wind-tunnel floor and carrying a weight to keep the spin axis vertical. The diameter of the largest wire was only ½ per cent of the diameter of the golf ball model. The two wires also served to supply a voltage to the motor inside the ball.

The upper support wire was attached to a strain-gauged arm which measured the lift force on the ball (with the vertical spin axis this was in fact a side force). The strain-gauged arm was in turn mounted on a rigid support attached to the wind-tunnel three-component balance. The wind-tunnel balance was used to measure the drag. The strain-gauged arm and the rigid support were shielded from the air stream by a streamlined fairing. The drag of the exposed wires was subtracted from the total measured drag.

The rotational speed of the ball was measured by using a stroboscope, which was regularly calibrated during the experiments. Where air speeds and rpm are quoted in the paper, these have been scaled to the appropriate values for a golf ball of 41.1 mm (1.62 in) diameter.

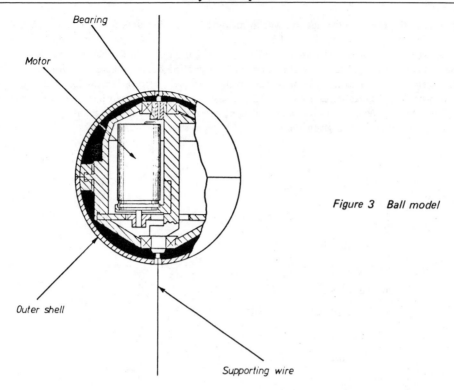

Figure 3   Ball model

2.2   EXPERIMENTAL RESULTS

Initially, in order to investigate the extent of support interference, measurements of C_D were made on a non-rotating smooth sphere. It is known (Goldstein[5] and Achenbach[6]) that the flow about a smooth sphere is highly dependent on Reynolds number and that, over a small Reynolds number range at about $Re = 4 \times 10^5$, the C_D reduces from about 0.45 to 0.1. In the high C_D regime the flow over the ball is laminar and separation occurs just ahead of the point of maximum thickness. The critical Reynolds number, i e the Reynolds number at which the rapid C_D reduction takes place, will occur earlier if the surface of the sphere is rough or the oncoming flow is turbulent. It was anticipated that if the support system seriously interfered with the flow then local regions of disturbed flow would be generated and the C_D would be reduced below the corresponding value given in the literature.

Measurements of C_D for a smooth, hexagonally-dimpled and conventionally-dimpled ball at zero spin are plotted in Figure 4 and it can be seen that the drag of the smooth ball remained high throughout the speed range.

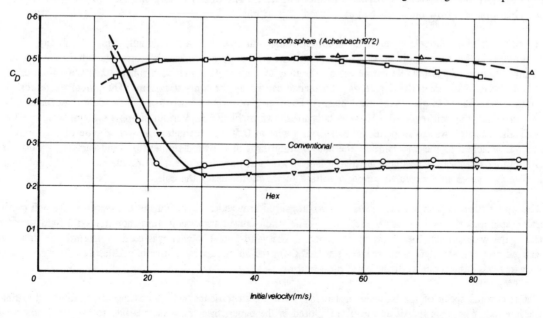

Figure 4   Drag coefficient versus wind speed for zero spin

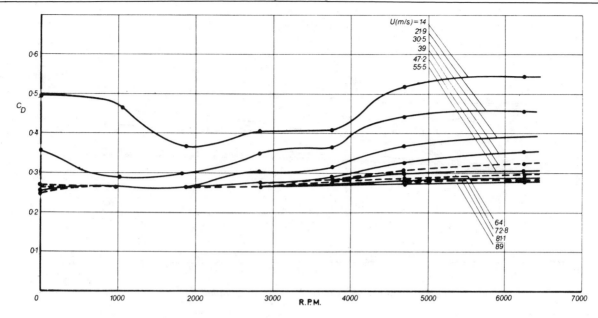

Figure 5   Drag coefficient.   Conventional ball

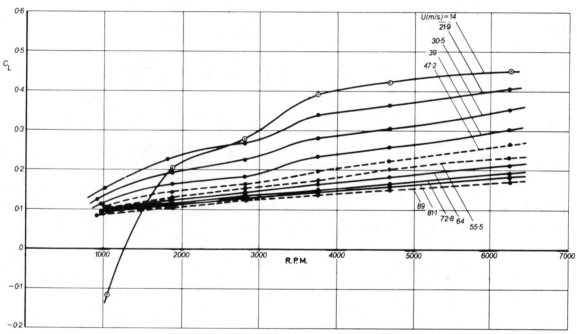

Figure 6   Lift coefficient.   Conventional ball

The smooth sphere results are shown to be close to the values of C_D measured by Achenbach[6] and they lie within the general scatter of published data. These results confirm that there was little support interference from the fine suspension wires over the Reynolds number range examined. Initial tests with thicker supports, however, did show a departure of the results from the published data. Clearly the dimpled balls show a critical Reynolds number behaviour and, although the C_D reduction is smaller than that occurring on a smooth sphere, it happens at a lower Reynolds number (see Figure 1). At speeds above about 30 m/s the hexagonally-dimpled ball has a slightly lower C_D than the conventional ball.

It is interesting to compare the variation with Reynolds number of the C_D of a conventionally-dimpled golf ball with the C_D of spheres roughened with sand grains. Achenbach[1] has carried out a series of measurements of C_D for spheres with various ratios of the average diameter of the sand grains k to the sphere diameter. Achenbach's results are shown in Figure 1; it can be seen that increase of k/d reduces the critical Reynolds number but that after the minimum the C_D rapidly rises, owing to forward movement of the transition point and the artificial thickening of the boundary layer by the roughness elements. It is difficult to compare dimples with sand roughness but presumably a relevant parameter is the ratio of the depth to the ball diameter (i e, $k/d \simeq 900 \times 10^{-5}$). Comparing golf ball C_D values with the results for sand-roughened spheres given in Figure 1, it can be seen that, at the same value of k/d, dimples are more effective at reducing the critical Reynolds number.

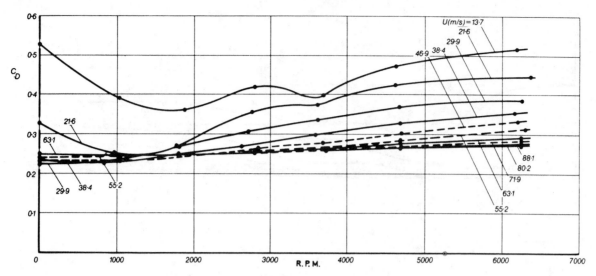

Figure 7   Drag coefficient. Hexagonally-dimpled ball

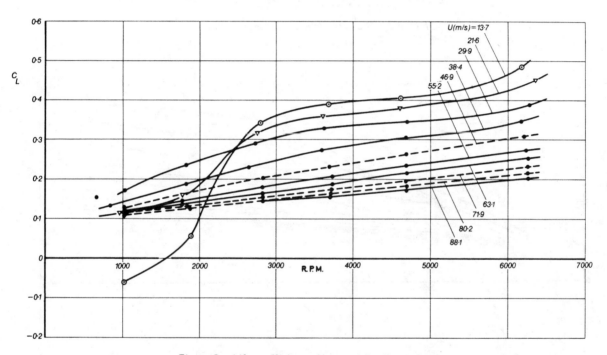

Figure 8   Lift coefficient. Hexagonally-dimpled ball

Another interesting point to note is that the golf ball C_D does not show the rapid increase with Reynolds number after the critical value. This suggests that the dimples are effective in tripping boundary layers without causing the thickening associated with positive roughness. It also suggests that dimples fix the transition point far forward. Clearly then the local flow about the dimples and the dimple shape must be important factors in determining C_D .

Measurements of the lift and drag on both hexagonally-dimpled and conventionally-dimpled balls were made at ten speeds between about 14 and 90 m/s and for six spin rates up to about 6250 rpm. This provides sets of data for the two balls over a wide range of Reynolds number and spin parameter. The variations of C_D and C_L with spin, for the conventional ball, are shown in Figures 5 and 6 respectively, for the ten values of velocity. The corresponding values for the hexagonally-dimpled ball are shown in Figures 7 and 8. In the absence of any gross Reynolds number effect, one would expect C_L to rise with increasing spin and, if one anticipates an induced drag effect similar to that observed with finite aspect ratio wings, C_D would rise. For a given rpm, increase of the velocity will reduce the spin parameter v/U and hence reduce the C_L and the C_D . The results in Figures 5 to 8 show these trends, apart from data taken at low speed at low rpm. In Figures 6 and 8 the lift is seen to be negative at the lowest speed below about 1500 rpm. At the lowest speeds the Reynolds number is below the critical value and the flow over the non-spinning ball is laminar. When the ball is spinning, disturbances first occur in the boundary layer passing over the part of the ball surface that is advancing against the main flow, i e, on what would be the

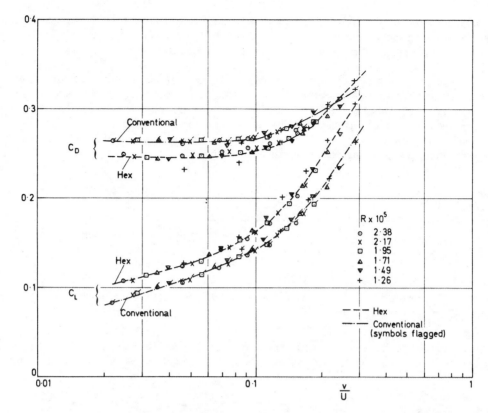

Figure 9   Comparison of conventional and hexagonally-dimpled ball.   Re = 1.3 x 10⁵ → 2.4 x 10⁵

underside of the ball. Transition occurs on this side, creating a negative lift force. With further increases in spin or speed, transition occurs on the top of the ball as well and the lift reverts to a positive value. This behaviour has been noted in the flow past rotating cylinders in the critical Reynolds number regime.

Above the critical Reynolds number, increases in rotational speed give a nearly linear increase in lift. However, it is interesting to note that, if the C_L curves are extrapolated back to zero spin, finite values of C_L ($\simeq 0.06$) are indicated rather than the zero value which is, of course, observed. The non-linear behaviour up to about 1000 rpm means that the lift can not be explained by any simple attached flow circulation theory. In order to compare the lift and drag developed by the two different dimple configurations, C_D and C_L values for each ball were plotted against the spin parameter v/U for different Reynolds numbers. The results are shown in Figure 9 for Reynolds numbers between 1.26×10^5 and 2.38×10^5; these correspond to speeds of between about 45 and 88 m/s. The collapse of data from different runs reflects the lack of dependence on Reynolds number and the overall accuracy of the experimental results.

At Reynolds numbers below 1.26×10^5 an effect of Reynolds number begins to appear and a collapse of data on a universal curve is not possible. It can be seen that, as the lift increases, the drag increases and thus a golf ball behaves in the same way as a more conventional lifting body where an induced drag increment is experienced. In general the hexagonally-dimpled ball is superior to the round-dimpled ball, having a higher C_L and a lower C_D, apart from the cross-over of the drag curves at high values of v/U, where C_L becomes very high.

3.   Computation of Trajectories

With knowledge of the aerodynamic forces and initial conditions it is possible to compute the complete trajectory for a golf ball drive. The equations of motion are

$$\ddot{x} = -\frac{\rho S}{2m}(\dot{x}^2 + \dot{y}^2)\left(C_D \cos\alpha + C_L \sin\alpha\right)$$

$$\ddot{y} = \frac{\rho S}{2m}(\dot{x}^2 + \dot{y}^2)(C_L \cos\alpha - C_D \sin\alpha) - g,$$

where x and y are measured in the horizontal and vertical directions respectively, g is the acceleration due to gravity, m is the mass of the ball and α is the inclination of the flight path to the horizontal, i e, $\alpha = \tan^{-1}(\dot{y}/\dot{x})$. These equations were solved using a step-by-step calculation procedure on a digital computer. All the wind-tunnel data were stored on the computer and at each time step the computer was programmed to interpolate the data to find the appropriate values of C_L and C_D. The step size chosen corresponded to 0.001 seconds of real flight time and test calculations using various time intervals indicated that the size adopted was sufficiently small and that the absolute error due to the numerical procedure was ±0.01 m on the prediction of range.

In order to compute the drive trajectory the initial spin, launch angle and launch velocity need to be known. A high-speed photographic technique was set up by Uniroyal Ltd to measure these parameters for a number of golfers. The golfers were classified into two groups; professional and amateur. The players repeatedly drove off, and on each occasion launch angle, spin and launch speed were measured. Mean launch parameters for professional and amateur golfers are shown in Table I. The results show that a typical professional drive, apart from being faster, is also at a lower angle and higher spin than a typical amateur drive.

TABLE I

	Spin (rpm)	Launch angle (degrees)	Launch velocity (m/s)
Professional drive	3450	6.1	68.1
Amateur drive	2450	9.9	56.7

A typical trajectory is shown in Figure 10. The aerodynamic forces lose importance as the flight progresses, since velocity is decreasing and in the second half of the trajectory the only significant force acting is that due to gravity.

Figure 10   Calculated trajectory . Initial conditions: velocity 57.9 m/s, elevation 10°, spin 3500 rpm

The amount by which the spin decays during a flight is not known accurately, although an estimate of the aerodynamic torque was obtained by measuring the electric current needed to rotate the ball in the wind tunnel. This indicated that the decay would be negligible during the part of the flight when the aerodynamic effects dominate. Various realistic spin decay assumptions, including the case where the decay is proportional to the spin squared, were used in the calculation of trajectory but very little dependence was found.

Some results of the computations for the effect of spin rate, initial velocity and initial angle are shown in Figures 11 to 13. Figure 11 shows the effect of spin for an initial velocity of 58 m/s and an initial elevation of 10° These results show that for these initial conditions maximum range is obtained for a spin of about 4000 rpm. Compared with a flight without spin, a spin of 4000 rpm gives about a 90 per cent increase in range. As a further

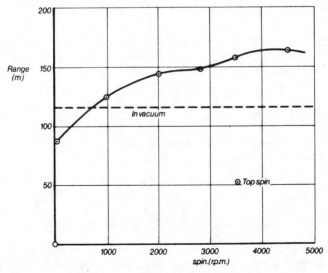

Figure 11   Effect of spin on range ·
Initial conditions: velocity 57.9 m/s, elevation 10°. Conventional ball

comparison the range for the same initial conditions, but without aerodynamic forces, is shown as the "in vacuum" result. It can be seen that above a quite modest spin rate the aerodynamic lift force helps to carry the ball farther than in the "in vacuum" case. To illustrate the importance of the aerodynamic forces still further, one result is shown in Figure 11 for a drive but with top spin. Changing the sign of the lift force reduces the range by about two thirds.

The effect of varying the initial velocity is shown in Figure 12. At low velocities very little lift is produced and the carry is small, but above 30 m/s the range rapidly increases, the dependence on changes in the initial velocity being almost linear. In practice, hitting the ball harder has the effect of increasing both the initial velocity and spin.

Figure 13 shows how the range is affected by the angle at which the ball leaves the tee. It should be noted that the curve will not pass through the origin since even at zero elevation a lift will be developed which will carry the ball an appreciable distance. The range is still increasing, albeit slightly, at an angle of 15°. This information could be used to indicate what initial spin and launch angle a club should be designed to impart to a ball in order to achieve maximum range. The measurement and computation technique developed could also be used to examine, say, the effects of head or tail winds and the trajectories of other golf shots, such as that from an iron where spin may be much higher.

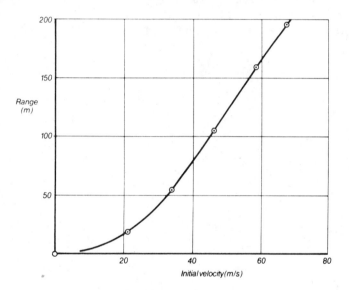

Figure 12   Range for varying initial velocity .

Initial conditions: elevation 10°, spin 3500 rpm. Conventional ball

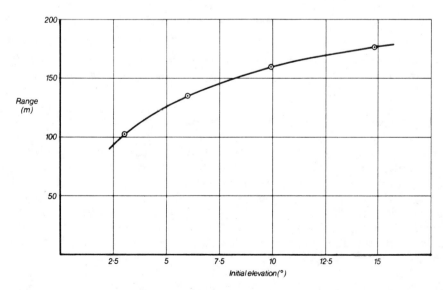

Figure 13   Effect of initial elevation on range .

Initial conditions: velocity 57.9 m/s, spin 3500 rpm. Conventional ball

4.   Comparison with Full Scale

The final phase of the investigation was to compare the ranges predicted from the wind-tunnel results with actual measured values using golf balls launched by a driving machine. By using a driving machine the launch conditions can be carefully controlled. These tests were performed by Uniroyal Ltd and we are grateful to them for the data supplied for use in this paper.

Their driving machine can be adjusted such that initial velocity, angle and spin can all be varied and in these tests it was set up to produce conditions typical of golf drives. It was not possible to monitor the complete trajectory but it was possible to mark the landing point and thus record the range. In order to make the comparison with the range computed using the wind-tunnel data, the high-speed photographic technique was again used to measure the launch parameters. Thus the aerodynamic performance is assessed independently of the mechanical properties of the ball.

Comparisons of the measured ranges with the computed ranges for a round-dimpled and hexagonally-dimpled golf ball are shown in Figures 14 and 15 respectively. Each measured range is an average range of nine balls driven twice; however, the scatter on ranges between individual balls was usually small. The corresponding predicted ranges were computed in each case. The agreement between measured and computed results was closest for the hexagonally-dimpled ball with, in most cases, the round-dimpled ball travelling slightly farther than predicted. For given initial conditions, this tends to exaggerate any predicted advantage of the hexagonally-dimpled ball. The driving machine results show the hexagonally-dimpled ball to travel approximately six yards farther than a round-dimpled ball under normal driving conditions.

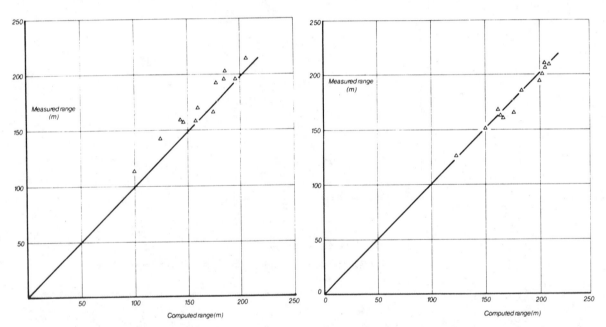

Figure 14   Comparison of measured and predicted range for a golf ball with round dimples

Figure 15   Comparison of measured and predicted range for a hexagonally-dimpled ball

5.   Discussion of Results

The aerodynamics of a golf ball depends on the flow induced by the dimples and unfortunately it is difficult to analyse this very closely, especially when the ball is spinning. Clearly they effect transition and reduce the critical Reynolds number below that for a smooth sphere. Unlike sand-roughened spheres, however, the C_D does not rise again immediately beyond the critical Reynolds number region. Probably sand roughness has the effect of thickening the turbulent boundary layer and thus making it less able to remain attached, whereas the dimples appear to act as more efficient trips. Hexagonal shaped dimples act as even more efficient trips than round dimples, perhaps by shedding into the boundary layer more discrete vortices from their straight edges. A detailed flow investigation of dimple flow is needed before more definite conclusions can be drawn.

Comparison of the measurements of C_D and C_L presented here with those of Davies[3], made at a Reynolds number of about 0.9×10^5, shows that he was in a regime where the flow was still weakly dependent on Reynolds number. His measurements of C_D are higher than ours whereas his measurements of C_L are slightly lower;

otherwise the trends are the same. Williams[4] deduces, from measured ranges, that $C_D = 151/V$, where V is the initial ball velocity in m/s, by assuming that flight does not depend on lift or spin. These assumptions are not supported by the findings of this research and his relation for C_D does not fit the data.

The superior aerodynamic performance of the hexagonally-dimpled ball suggests that, for the same launch conditions, it should travel farther than a round-dimpled ball. This was borne out by driving machine tests, although the advantage of the hexagonally-dimpled ball was not as great as that suggested by the wind-tunnel experiments. This was due to an under-prediction of range for the conventional ball, the predictions for the hexagonally-dimpled ball being very close to the measurements. Both balls were constructed using golf ball mould drawings and no attempt was made to model the coating applied to a finished ball. This will mean that the dimple edge radius may not be truly scaled and this may be an important parameter in determining C_D and C_L. More work is required to examine the effect of edge radius on the aerodynamic performance.

6. Conclusions

A wind tunnel has been used to measure the lift and drag acting on golf ball models over a wide range of Reynolds number and spin rate. The effect of changing the dimple shape from round to hexagonal has been examined. The dimples trip the boundary layer and induce an earlier critical Reynolds number. The hexagonally-dimpled ball showed a superior aerodynamic performance compared with a round-dimpled ball.

Using the aerodynamic data, ranges were computed for various initial conditions and the effects of initial velocity, angle and spin were investigated separately. Comparisons with full-scale data from a driving machine showed that the hexagonally-dimpled ball range was closely predicted whereas the round-dimpled ball range tended to be slightly under-predicted, possibly owing to minor model inaccuracies. With the data presented in this paper the influence of the aerodynamic forces on a wide range of golf shots can be analysed.

This investigation has, at the same time, highlighted the effectiveness of dimples as a method of causing early boundary-layer transition without the drag penalties associated with other generally accepted trips which protrude from the surface.

Acknowledgements

The authors wish to acknowledge the help given by Messrs Uniroyal Ltd, and in particular by Dr T Moore, who very kindly provided us with the range data referred to in this paper. We are also indebted to Dr M E Davies for his assistance with the computer calculations.

References

1.   E Achenbach — The effects of surface roughness and tunnel blockage on the flow past spheres. *Journal of Fluid Mechanics*, Vol 65, p 113, 1974.

2.   J Maccoll — Aerodynamics of a spinning sphere. *Journal of the Royal Aeronautical Society*, Vol 32, p 777, 1928.

3.   J M Davies — The aerodynamics of golf balls. *Journal of Applied Physics*, Vol 20, p 821, 1949.

4.   D Williams — Drag force on a golf ball in flight and its practical significance. *Quarterly Journal of Mechanics and Applied Mathematics*, Vol XII, p 387, 1959.

5.   S Goldstein (Editor) — *Modern Developments in Fluid Dynamics*, p 16. Clarendon Press, Oxford, 1938.

6.   E Achenbach — Experiments on the flow past spheres at very high Reynolds numbers. *Journal of Fluid Mechanics*, Vol 54, p 565, 1972.

Do springboard divers violate angular momentum conservation?

Cliff Frohlich

Marine Science Institute, University of Texas, 700 The Strand, Galveston, Texas 77550
(Received 15 November 1978; accepted 26 February 1979)

No. However, divers and trampolinists can perform somersaults and twists even though they have zero angular momentum at all times during the stunt. Also, if a diver is somersaulting in space and possesses angular momentum only about his somersaulting axis, he can make a single discrete change in the position of his arms which initiates continuous twisting motion even in the absence of any applied torque. These apparent paradoxes have confused both physicists and coaches for some time. The present paper attempts to reduce this confusion. It discusses several different methods that performers use to initiate somersaults and twists and presents concrete examples of each method. Wherever possible quantitative calculations are presented and evaluated using information about the moments of inertia, mass, etc., of "typical" performers.

Reprinted from *American Journal of Physics*, **47**, 7. ©1979 American Association of Physics Teachers.

I. INTRODUCTION

Divers, trampolinists, and gymnasts can perform rotations of their bodies of several different kinds. The first kind is the *somersault,* which is a rotation of the body about an axis along a line going from the performer's left side through his center of mass to his right side. In Fig. 1, the axis of rotation for somersaulting motion is marked with an "*S*" for a performer in several common diving positions. The second basic kind of rotation is the *twist,* which is a rotation of the body about an axis going from the performer's head through his center of mass to his feet. In Fig. 1, this axis is marked with a "*T*" for the diver in positions *A, B, C, D,* and *E.* Twists are usually performed with the body straight, and with no bend in the knees or at the waist. Only a tiny fraction of all stunts consist principally of a rotation about the third possible axis, which is along a line going from directly in front of the performer through his center of mass, and out through the center of his back. However, this paper will show that small rotations about this axis do play an important role in the initiation of some kinds of twisting stunts, particularly twisting somersaults.

Apparently very few physicists have observed springboard divers or trampolinists closely. Recently, all the graduate students, postdoctorates, and faculty in the Physics Department at Cornell University were given a questionnaire with specific multiple choice questions about the physical possibility of performing certain somersaulting and twisting stunts (see Fig. 2). One question asked if it is possible for a trampolinist with no angular momentum to execute a quarter somersault (a 90° rotation about his left-to-right axis). Another question asked if a somersaulting springboard diver could initiate twisting motion without any torques being applied to his body, that is, beginning the twist only *after* he is no longer in contact with the diving board. As we show later in this article, the answer is that trained divers can perform both of these motions without violating angular momentum conservation, Nevertheless, of the 59 physicists who responded to the questionnaire (see Table I), 34% incorrectly answered the first question, and 56% incorrectly answered the second—an astonishingly high proportion of misses for multiple choice questions.

These results illustrate the need for a paper discussing the basic physics of somersaulting and twisting stunts, particularly since there are at present no physics books or articles which treat these phenomena in any detail. Furthermore, divers, trampolinists, and gymnasts do provide excellent illustrations of certain peculiar features of the free rotation of rigid and semirigid bodies. Since more than 300 colleges and universities in the U.S. possess competitive swimming teams,[1] these phenomena can be observed by physicists and physics students on almost any campus.

Coaches and physical educators (as opposed to physics educators) also may find a review of the physics of diving to be useful. Several books written by or for coaches discuss somersaulting and twisting, and exhibit varying degrees of insight and/or confusion about the physics processes that occur.[2-5] Although the author of this paper is not qualified to write a "how-to" paper for performing stunts, the present work does provide the most precise account available on the physics of somersaulting and twisting stunts.

Historically, the earliest research relevant to these problems was not concerned with the motions of the human body, but instead with the twisting of cats, who land on their feet when dropped upside down from a height of a meter or so.[6-9] More recently, the possibility of space exploration has caused more quantitative work to be done.[10,11] Although Smith and Kane[10] were primarily concerned with motions of men in space, their paper also includes an excellent comprehensive review of much of the earlier work. In other related work, high-speed cameras have been used to make detailed observations of an athlete performing a back somersault with a twist[12] and a computer has been used to simulate the motions that occur in somersaults without twists.[13] However, neither of these studies of athletes discussed the physics associated with these stunts in any detail.

Because divers, coaches, astronauts, and physicists commonly use different terminology to describe motions of the human body, a quick summary of some synonymous terms is given here to minimize confusion. For example, the *left-to-right axis* is sometimes called the transverse axis, and somersaulting motion about this axis may be called "motion in the sagittal plane." Similarly, the *head-to-toe* axis is also called the longitudinal axis or the long axis, and twisting motion is said to take place "in the transverse plane." Finally the *front-to-back axis* is also called the

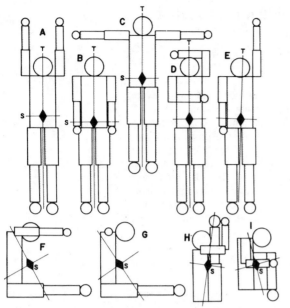

Fig. 1. Configuration of body segments of "typical" man in nine positions commonly used by divers and trampolinists. For each position, the solid diamond is at the center of mass, with the long axis of the diamond oriented along the principal axis associated with the principal moment of inertia I_1 in Table III. The axis of rotation for somersaulting dives is marked with an "S." and the axis of rotation for twisting is marked with a "T." The nine positions are; A: Layout throw; B: Layout somersault; C: Pretwist layout; D: Twist position; E: Twist throw; F: Loose pike; G: Pretwist pike; H: Tight pike; and I: Tuck.

medial axis, and rotations about this axis occur "in the frontal plane." When more than one term could be used to describe a stunt, in this paper the term which appears in italics in this section will be used.

Divers perform somersaulting rotations in several positions, for example, the *tuck* position (knees and waist bent, as in I in Fig. 1), the *pike* position (knees straight but waist bent, as in F, G, or H in Fig. 1), and the *layout* position (knees straight and waist not bent, A, B, and C in Fig. 1). When both somersaulting and twisting motions occur in the same dive, the rules place no restrictions on bending the knees and waist, and hence the dive is performed in the so-called *free* position (D and E in Fig. 1 are among the typical configurations assumed during twisting somersaults). Physicists commonly specify the direction of rotations using the right hand rule. When performing a *forward* somersault, a diver leaves the diving board facing the water and his rotation direction along his left-to-right axis is to his left. In a *backward* somersault he leaves the diving board with his back to the water and the direction of rotation along his left-to-right axis is to his right. For reference, Batterman[3] has published in paperback a collection of diving pictures

Table I. The fraction of respondents to the questionnaire who incorrectly answered the questions posed in Fig. 2. Eight diving coaches and 59 physicists responded to the questionnaire. The last column is the fraction of incorrect answers expected if the respondents had chosen their answers randomly.

	Coaches	Physicists	Random guess
Question 1	0.38	0.34	0.50
Question 2	0.13	0.56	0.67

which includes most of the dives commonly performed in collegiate competition.

II. PHYSICAL CHARACTERISTICS OF DIVES AND DIVERS

Some stunts are "possible" because no law of physics prevents their performance, but are "impossible" in practice because of the limitations of their performers. For example, ballistics tells us it is possible for a cow to jump over the moon, but we know that cows can not jump that high. In order to evaluate in terms of physics what occurs in diving and trampolining stunts, it is useful to have numerical estimates of certain characteristics of typical performers executing typical stunts.

A. Masses of body segments

The relative masses of body segments in Table II have been taken from a standard physical education text by Dyson,[2] and then rounded slightly for convenience in the calculations of the moments of inertia. The classic study of relative masses of body segments was done by Dempster,[14] and his results have been summarized elsewhere by several authors.[15] Dempster's source of data consisted of cadavers

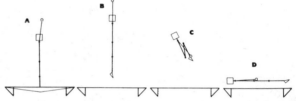

Fig. 2. *Question 1:* The stick-figure above is performing a simple trampoline stunt known as a "back-drop." In this stunt, he bounces, tucks, and lands on his back on the trampoline bed. *Question 1 in divers language:* Could a trained performer complete a back drop without any "throw" from the bed of the tramp? In other words, if the body of the trampolinist is not rotating at all at point "B" (after his feet have left the tramp bed) can be still do a back drop? *Question 1 in physicists' language:* If the trampoinist has zero angular momentum at "B" is it possible for him to land on his back as in "D"? Check one answer: □ Yes, a trained trampolinist could successfully complete the stunt as shown. □ No, even a well-trained individual could not do that stung if at "B" he had no rotating motion. *Question 2:* One dive that is listed in the AAU's *Official Diving Rules 1978* for a three meter springboard is a forward one and one half somersaults with three twists. Although this is a very difficult dive, there are probably more than a hundred divers in the United States who can do it. *Question 2 in divers' language:* Suppose the diver throws the somersault, but does not begin to throw the twists after his feet have left the board. Is it possible for him to do multiple twist dives like the triple twister even though he does not throw the twist until after he leaves the board? With practice, could a diver learn to throw twists in either direction, even though he did not begin the twist on the board? *Question 2 in physicists' language:* If the diver has initiated only somersaulting motion on the board, is it possible for him to initiate the twists after his feet have left the board? Could he choose to twist either to the left or to the right, depending on how he threw the twist after he left the board? Check one answer: □ Yes, a diver really can do a multiple twist dive such as a triple-twisting one and one-half somersault, and if properly trained he could choose the twist direction after his feet left the board. □ Yes, trained divers can do multiple twist dives where they initiate the twist after they leave the board. However, no matter how hard they train the direction of the twist is already determined before their feet leave the board. □ No, even a well-trained diver could only do a multiple twist dive if he initiated the twisting motion while his feet still touched the board.

Table II. Body segment masses and shapes for "typical" man with total mass of 75.64 kg and height 1.82 m.

Segment	Mass (Kg)	Description (cm)
Head	5.575	sphere: $r = 11$
Trunk	32.400	rectangular solid; $60 \times 30 \times 18$
Upper arms (2)	2.356	cylinder; $r = 5$, $h = 30$
Forearms (2)	1.781	cylinder; $r = 4.5$, $h = 28$
Hands (2)	0.523	sphere; $r = 5$
Thighs (2)	8.650	cylinder; $r = 8$, $h = 43$
Lower legs (2)	4.086	cylinder; $r = 5.5$, $h = 43$
Feet (2)	1.436	sphere; $r = 7$

of eight subjects who died in middle or old age, and as such his results are not typical for divers and trampolinists. The relative masses reported in Table II are within the ranges reported by Dempster,[14] but on the average a lower percentage of the body weight is concentrated in the trunk, and a higher percentage in the limbs.

B. Principal moments of inertia

The moments of inertia reported in Table III are calculations of the author, as no detailed moment of inertia calculations for most diving body positions could be found in the literature. To make the calculations, the author divided the body into 14 segments and assigned each segment an approximate convenient shape which had an easily calculable moment of inertia (e.g., sphere for head, rectangular solid for trunk, cylinders for legs, etc.; see Table II). The principal dimensions of each segment were determined by measuring the appropriate dimensions of the author's own body with a meter stick. The remaining dimension (e.g., the radius of the leg cylinders) was chosen so that the conveniently shaped segment would have the density of water and the mass reported in Table II. Finally, a computer program calculated the center of mass and principal moments of inertia for typical positions used while somersaulting and twisting (Table III and Fig. 1).

For a few of the body positions it is possible to compare these results to other work. The calculations reported in Table III fall within the range of values calculated by Hanavan[11] for men of comparable height in body positions similar to B, C, F, and G in Fig. 1. Hanavan's[11] calculations extended the work of Santschi et al.[16] They measured the moments of inertia for 66 subjects in several body positions by the compound pendulum method, but had made no attempt to correct their measurements to obtain values for the moments along the principal axes. Rackham[4] reported that the ratio of moments for divers somersaulting about the left-right axis in the tuck, pike, and layout positions is approximately 1:2:4, also in substantial agreement with the results in Table III.

C. Stability

For a rigid body with three distinct moments of inertia, stable rotational motion is possible only about the principal axes associated with the largest and the smallest moments of inertia.[17] Rotation about the third principal axis is only metastable; the smallest perturbations of the rotation grow to become mysterious "wobbles" or twists.

When divers and trampolinists perform somersaults, are they rotating about a stable axis? Somersaults are rotations about the left-right axis, and Table III suggests that a performer somersaulting in a tuck or loose pike position is somersaulting about the axis associated with his largest

moment of inertia, and thus the motion is stable. The moment of inertia for somersaulting in a tight pike position (Table III) is very close to the largest moment of inertia. Since few real divers can pike as tightly as the "typical man" in Fig. 1, probably the largest moment of inertia for piked real divers is associated with the somersaulting axis, and so this motion is stable. On the other hand, for a layout position, the largest moment of inertia is definitely not associated with the somersaulting axis, and so the somersaulting motion is not a stable one. The fact that the layout position is unstable for somersaulting may be partly responsible for the "side cast" or partial twist of the body which is sometimes observed as a diver enters the pool. This is a common problem for divers performing layout somersaults.

D. Time

Time is the fundamental constraint that limits all divers and trampolinists from performing more difficult stunts. A diver leaving a board is a projectile—if he simply drops from a 1-m board, 0.45 sec will elapse before he reaches the water. Similarly, 0.8 sec will elapse if he drops from a 3-m board. However, if he can use the board to raise his center of mass 2.0 m, he has about 1.5 sec to complete a dive from the 1-m board, and about 1.7 sec from a 3-m board. This additional 0.2 sec permits the performance of more complicated motions from the 3-m board motions with an additional twist or an additional somersault. For example, in American Athletic Union (AAU) sanctioned competition divers can perform forward one and one-half somersaults with four twists from the 3-m board, but with no more than three twists from the 1-m board.[18]

Clearly, a key element in any diver's technique is to achieve the maximum height possible so as to have sufficient time to complete complicated stunts. However, since this paper is concerned primarily with somersaulting and twisting, the physics of achieving height will not be discussed.

III. INITIATING SOMERSAULTS AND TWISTS

Somersaults and twists can be classified in more than one way. In this section, they are discussed in terms of the forces and torques acting to initiate the body rotation. In these terms, there are three types of somersaults, and three types of twists.

Table III. Principal moments of inertia for "typical" man in positions shown in Fig. 1. I_1 is the principal moment for the axis with direction closest to that of the man's spine (twist axis or head-to-toe axis), I_2 is the principal moment along a left-to-right axis (somersault axis), and I_3 is the principal moment along an axis closest to the front-to-back axis. Units are kg m^2. Letters beneath "Figure 1" column refer to appropriate diagram in Fig. 1.

Figure 1	Description	I_1	I_2	I_3
A	Layout throw	1.10	19.85	20.66
B	Layout sault	1.10	14.75	15.56
C	Pretwist layout	3.42	16.38	19.17
D	Twist position	1.06	16.65	17.24
E	Twist throw	1.08	17.41	18.20
F	Loose pike	4.83	10.45	7.53
G	Pretwist pike	5.54	10.09	10.42
H	Tight pike	1.75	5.89	6.05
I	Tuck	20.3	3.79	3.62

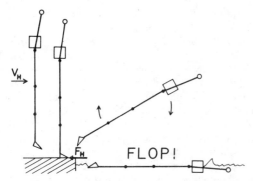

Fig. 3. Somersaulting can be initiated by horizontal forces even if the performer does not "throw" with his arms. In a "front header" from the side of the pool the diver runs and jumps into the air so that his body is moving forward with no rotational motion, but with only translational velocity (stick figure at far left). However, when his feet contact the pool deck, the pool deck exerts a horizontal impulsive force F_H on his feet, slowing his translational velocity and creating a torque that tends to initiate rotation. Although this dive can be observed at any neighborhood swimming pool, in competition the forward dive is generally not performed in this way.

A. Somersaults

(*i*) *Torque for somersault due to horizontal forces.* The most common way to initiate somersaulting motion is to apply torques to the body by applying horizontal forces to the feet. This can be done in more than one way. For example, if the diver's body is moving forward as he reaches the end of a diving board, the board applies horizontal forces to his feet to initiate a forward somersault (Fig. 3). Somersaulting motion will occur even if he never "throws" the somersault with his arms, shoulders, and head.

A much more controlled somersault is possible if the individual is allowed to "throw" with his upper body. As an example, the left picture in Fig. 4 shows how a forward somersault generally is initiated on a trampoline. "Throwing" the arms and head forward and down involves a rotation of the arms, which possess angular momentum. In free space, the lower body would rotate in the opposite direction (since angular momentum is conserved). However, since the trampolinists's feet are in contact with the trampoline, the trampoline exerts a force on the trampolinist which resists this opposite rotation, and provides the torque which begins the somersaulting rotation. The harder a performer "throws" with his upper body, the larger the horizontal force will be providing the torque, and the more angular momentum his body will have when the vertical forces provided by the trampoline cause his body to become airborne. The individual also can minimize or control the horizontal motion that results from these horizontal forces by making subtle adjustments in the position of his center of mass while his feet still touch the trampoline. This method of initiating somersaults—with torque provided by horizontal forces that counteract a rotation or "throw" of a portion of the body—is by far the most important method for trampolinists and divers.

In spite of the important role that horizontal forces play to initiate somersaulting, few divers or trampolinsts are aware that horizontal forces are being applied to their feet. Instead their attention is focused on how they "throw" with the upper body, and on their "balance" (the position of their center of mass). A brief calculation shows that this is be-

cause the magnitude of the horizontal forces that produce the torque is considerably smaller than the magnitude of the vertical forces that make the performer airborne. For example, consider a person performing a single somersault on the trampoline. If F_v and F_h are the vertical and horizontal forces acting for a time T on a performer's feet, then the initial takeoff velocity V_i and the initial rotational velocity ω_i are given by

$$V_i = F_v T/M \text{ and } \omega_i = F_h TL/I_{\text{lay}}, \qquad (1)$$

where M is his mass, I_{lay} is his moment of inertia about his left-right axis, and L is the distance from his feet to his center of mass while the forces are being applied by the trampoline: To simplify the calculation, assume his body is rigid and ignore the angular momentum possessed by his upper body because of the "throw." If he tucks his body at the moment of takeoff, his rotational velocity will increase by a factor of $I_{\text{lay}}/I_{\text{tuck}}$, where I_{tuck} is his moment of inertia in the tuck position, and so

$$\omega = F_h TL/I_{\text{tuck}}. \qquad (2)$$

When his center of mass is at the highest point, he will have completed one half somersault. The elapsed time T_e after takeoff will be

$$T_e = V_i/g = \pi/\omega \qquad (3)$$

and his height H above the trampoline will be

$$H = (1/2)gT_e^2 = V_i\pi/2\omega. \qquad (4)$$

Substituting in for V_i and ω from (1) and (2), one finds

$$F_h/F_v = \pi I_{\text{tuck}}/2MLH. \qquad (5)$$

If $L = H = 1$ m, and I_{tuck} and M are as in Tables II and III, then $F_h/F_v = 0.08$.

(*ii*) *Torque for somersault due to vertical forces.* An individual can perform a somersault without any throw from the diving board or trampoline, even if only vertical forces act on his feet. In this case, his center of mass cannot be directly above his feet, and it is because he leans either forward or backward that the vertical forces produce a torque (see Fig. 5). This method of producing somersaulting motion is seldom employed on the trampoline, as it is difficult for the trampolinist to control the amount of lean, and

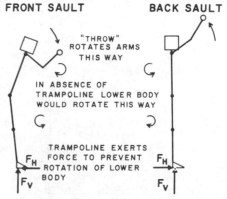

Fig. 4. Somersaulting can be initiated by a "throw" of the upper body while the performer's feet are still in contact with a trampoline or diving board. The throw creates horizontal forces, producing torques which initiate rotation. The direction of the rotation is the same as the direction of the throw.

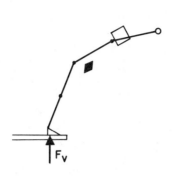

Fig. 5. Somersaulting can be initiated by vertical forces, even if the performer does not "throw" with his arms. In the picture the diving board exerts an upward force on the diver which creates a torque about his center of mass (solid diamond) causing him to rotate forward. In practice both front and back somersaults can be initiated in this manner if the diver "leans" as the board exerts a vertical force. The lean is exaggerated in the picture above.

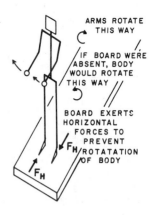

Fig. 7. Torque twist can be initiated by "throwing" the arms in the direction of the twist. If the performer's feet are in contact with the diving board, trampoline, or floor, then when the performer jumps into the air while throwing his arms in this manner, the horizontal forces create torques which cause him to twist.

a great deal of unwanted horizontal motion or "travel" often results. The lean also makes it more difficult to get enough height to perform difficult stunts.

Nevertheless, torques arising from vertical forces are important for a few kinds of stunts, particularly for divers performing multiple forward somersaults. For example, a slight lean is useful for divers performing forward two and one-half somersaults and forward three and one-half somersaults where a great deal of angular momentum is needed. In addition to the angular momentum resulting from the vertical forces, in practice the slight lean permits the diver to "throw" slightly earlier and slightly harder, which also helps to increase his angular momentum.

(*iii*) *Torque-free somersaults.* A limited amount of body rotation can be accomplished in the complete absence of any torque whatsoever. For example, if a trampolinist with legs and body straight "windmills" his arms forward about his shoulders, his entire body will tilt backwards. Because the moment of inertia of his arms is much smaller than the

moment of inertia of his body, a 360° rotation of his arms produces about a 20° rotation of his body. Thus, even though his angular momentum is exactly zero throughout the entire stunt, he has rotated or "somersaulted" 20°.

Figure 6 illustrates that it is possible to perform a torque-free somersault with nearly 90° rotation even without obviously "windmilling" the arms. Formulas used in the calculation of the amount of rotation drawn in Fig. 6 are derived in the Appendix. In practice, torque-free somersaults are not the essential part of any usual stunt, with the possible exception of the tuck back drop on the trampoline (a 90° back somersault, illustrated in Fig. 6). However, the same principles do play a small part in the execution of the ending of certain stunts, e.g., allowing a diver performing tucked somersaults a certain amount of control over his body's angle of entry into the water.

B. Twists

(*i*) *Torque twists.* Conceptually, the simplest way to initiate a twist is to apply a torque to the performer's body. For example, even nonathletic individuals standing on a hard surface generally can jump into the air and twist 180° or more by "pushing off" from the floor with their feet. Similarly, divers and trampolinists can initiate twists by pushing with their feet against the diving board or against the trampoline.

As with somersaults, the forces that initiate the rotation can be controlled most easily if the performer "throws" his arms in the direction of the twist *before* his feet lose contact with the floor. In the absence of forces, his lower body would rotate in the opposite direction from his arms. However, since the floor can apply forces to his feet these forces prevent this rotation in the opposite direction, and tend to cause his body to rotate in the same direction that he threw his arms (Fig. 7).

(*ii*) *Torque-free twists with angular momentum.* No torque whatsoever is applied to initiate twisting in the most important and most common method of initiating twists during somersaulting stunts. This mechanism can best be appreciated by carefully studying an example (Fig. 8). Suppose that a diver has initiated stable somersaulting motion and is in the layout position. Suppose further that he is no longer touching the diving board, that his body possesses considerable angular momentum about his left-to-right axis, and that his arms are extended to his sides as in Fig. 8(a) and in C in Fig. 1. Now the diver suddenly "throws" his right arm above his head and his left arm down to his side as in E in Fig. 1, moving his arms in the plane of

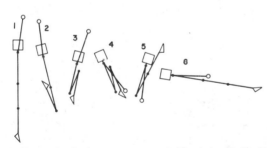

Fig. 6. Torque-free back quarter somersault ("back-drop"). If a "back drop" is executed in discrete steps as shown, the body will have rotated a total of 82° between the first and the sixth step, even though his angular momentum is exactly zero at all times. If the performer wished to finish the back drop with his body position exactly similar to his position in the first diagram but rotated 82°, after the sixth diagram he could raise his arms in the lateral plane of his body (along his sides). In principle a trampolinist could also rotate 82° forward by performing the steps shown in the reverse order, or 6-5-4-3-2-1. For each diagram, the amount of rotation shown was calculated as described in the Appendix, with the body consisting of two rigid pieces connected by a hinge. Each rigid piece is built from segments having the masses and intrinsic moments of inertia of the body parts listed in Table II. In practice, back drops are seldom done exactly as indicated in this figure. For example, usually steps two and three would be performed simultaneously, with the legs being tucked up in one motion. Also, in practice the neck and trunk are somewhat flexible, allowing a performer with zero angular momentum to rotate more than 82°.

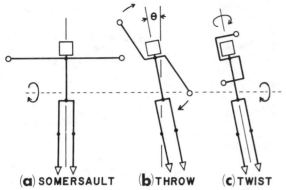

(a) SOMERSAULT **(b) THROW** **(c) TWIST**

Fig. 8. Diver possessing angular momentum can initiate twisting even in the absence of any external torques. In (a) above, a somersaulting diver has angular momentum only about his left-right axis (dotted line) and has no twisting motion. At the instant pictured in (b) he sharply "throws" his left arm down and his right arm up laterally in the plane of his body. Because he has no angular momentum about his front-back axis (normal to the plane of the picture) as his arms rotate clockwise, his body rotates an angle θ counterclockwise. However, this causes his body to begin a continuous tumbling or twisting motion about his long axis in order to conserve angular momentum (see Fig. 9). In (c) this twisting motion as well as his somersaulting motion continue even when he is no longer "throwing" with his arms, but is in effect a rigid body with no external forces acting upon him.

his body as in Fig. 8(b). Viewed from the front, his two arms effectively are rotating clockwise, and so his lower body must react by rotating counterclockwise an angle θ about his front-to-back axis. Because of this rotation of θ of his body axes, his left-to-right body axis is no longer aligned with his angular momentum axis, and so to conserve angular momentum his body begins to tumble or twist about his head-to-toe axis. His body will continue to twist with considerable rotational speed even after his "throw" is complete, and even if he is holding his body rigidly.

The cause of this torque-free twisting is particularly clear if one considers the diver's moment of inertia tensor **I** in an inertial reference frame. Suppose the inertial reference frame is chosen so that just before the "throw" [Fig. 8(a)] the diver's principal axes are along the axes of the inertial frame, and thus **I** possesses no off-diagonal components. His angular momentum vector **H** and angular velocity vector ω also are parallel to one of the coordinate axes, e.g., call it the $(0, 1, 0)$ axis. Immediately after the throw [Fig. 8(b)] the diver's principal axes are no longer along the axes of the inertial coordinate system, and thus **I** now possesses off-diagonal components. Because of these off-diagonal components, ω must possess components along the $(1, 0, 0)$ axis and/or the $(0, 0, 1)$ axis in order to ensure that the product **I·ω** = **H** remains fixed in space along the $(0, 1, 0)$ axis.

With only a few assumptions one can derive an extremely simple expression for the rate of twisting per somersault for torque-free twists with angular momentum. First assume that the "throw" of the arms [occurring in Figure 8(b)] is fast enough so that no twist about the diver's head-to-toe axis occurs during the throw (e.g., in a time small enough so that a negligible amount of somersaulting takes place). Second, assume that after the "throw" is complete, two of the diver's principal moments of inertia are equal. Although this is not strictly true, in a layout position the two largest principal moments are close enough to one another (Table

III) so that this assumption should give a good approximation of the twisting motion.[19]

With these assumptions, the diver is a rigid symmetrical top experiencing force-free motion which can be described completely in terms of simple trigonometric functions by solving Euler's equations of motion.[17] His rate of twisting is the angular velocity ω_t (total spin) about the body axis associated with his smallest principal moment of inertia (the long axis), and his rate of somersaulting is the rate of free precession ω_s that his angular velocity vector experiences about the angular momentum axis. In particular, if I_1 and I_2 are the two principal moments of inertia of the twisting diver, and if $\omega_1 = \omega_t$ and ω_2 are his rotational velocities about these principal axes, then Fig. 9 shows that

$$I_1\omega_t/I_2\omega_2 = \tan\theta \text{ and } \omega_2 = \omega_s \cos\theta. \tag{6}$$

Combining, we see that the rate of twisting per somersault is

$$\omega_t/\omega_s = I_2 \sin\theta/I_1. \tag{7}$$

This is the fundamental equation that governs the number of twists per somersault that a diver can perform if he initiates the twisting entirely after his feet have left the board. Note that the direction of the twists depends solely on whether the diver throws his arms clockwise or counterclockwise. Note also that the rate of twisting depends critically on the ratio of the principal moments of inertia (so that "skinny divers twist easier") and also on the angle θ that his body goes off axis when he "throws" his arms (Fig. 8). Practical ways of making this rate as large as possible will be discussed in the last section of this paper.

It is worth emphasizing that the kind of "throw" that initiates torque twists is quite different from the throw that initiates torque-free twists with angular momentum. For torque twists, the arms are thrown so that they effectively

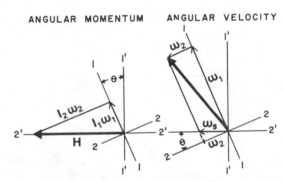

Fig. 9. Vector diagrams of the angular momentum vector and angular velocity vector in the inertial frame the instant after the twist is thrown. In both diagrams, the performer's long axis (twist axis) is along the 1' axis before he "throws" his twist, but along the 1 axis after the twist begins. The performer is somersaulting about the 2' axis, along which angular momentum **H** (heavy arrow in left diagram) is conserved. Before the twist is thrown, there is no component of angular momentum about the performer's long axis, and so the performer's angular velocity vector is along the 2' axis, parallel to the angular momentum vector. After the twist is thrown, there is a component $I_1\omega_1$ of angular momentum along the 1 axis, and so twisting rotation begins. The vector components of angular momentum along the 1 and 2 axes satisfy $I_1\omega_1/I_2\omega_2 = \tan\theta$. The angular velocity vector is no longer along the 2' axis. The diagram on the right shows that ω_s, the component of the angular velocity vector (heavy arrow) along the 2' axis (somersaulting axis) is related to ω_2 by the relation $\omega_s \cos\theta = \omega_2$.

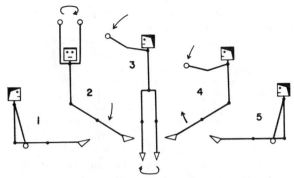

Fig. 10. "Swivel hips," a simple example of a trampoline stunt in which a performer can change the relative moments of inertia of his upper and lower body to accomplish a torque-free twist. The performer bounces from a sitting position (1), and begins to raise his arms over his head by bringing them forward and up as shown by the arrow. In (2), he rotates his upper body to turn his arms and shoulders as far around as possible, and he beings to drop his legs. Because his legs are piked and his arms are raised, the moment of inertia of his lower body is large relative to the amount of inertia of his upper body, and thus his legs experience little rotation. In (3) he brings his arms forward and down and "swivels" his hips beneath him while his arms are extended. In this step the moment of inertia of his lower body is smaller than that of his upper body. In (4) he brings his feet and legs up into a pike, and in (5) the stunt ends as it began, in a sitting position.

rotate about the body's head-to-toe axis, thus producing torques about this axis. For torque-free twists with angular momentum, the arms are thrown so that they rotate the body's principal axes relative to the angular momentum vector. In fact any body motion made during the dive will initiate twisting if the effect of the body motion is to reduce the angle between the head-to-toe axis and the angular momentum vector.

(*iii*) *Torque-free twists with zero angular momentum* or "cat twists." It is possible to rotate the body about the head-to-toe axis even though the body has no angular momentum and even though no external torques are applied. Like torque-free somersaults, these twists do not provide continuous twisting motion, but easily can produce rotations of 180° with only a few simple motions.

For example, if a cat is held upside down by its four legs and dropped without any initial rotation from a height greater than about half a meter, the cat will perform a sequence of motions that varies the relative moments of inertia of the front and rear portions of its body so as to cause it to land on its feet.[6,7] On the trampoline, a simple stunt that utilizes this type of twisting is called "swivel hips." One way of doing this stunt is described in Fig. 10.

It is also possible to perform torque-free twists without varying the relative moments of inertia of different portions of the body.[10] Figure 11 describes a simple body which demonstrates this. It consists of two identical plumb-bob shaped objects which are in contact along their conical ends. The three principal moments of inertia of each object are I, I, and J. The two objects roll without slipping on their conical ends with angular velocity ω, and simultaneously rotate with angular velocity Ω about the axis bb' (see Fig. 11). The angular momenta $\mathbf{H}_T$ and $\mathbf{H}_B$ of the two bodies are

$$\mathbf{H}_T = J\omega\mathbf{j}_T - J\Omega\cos\alpha\mathbf{j}_T + I\Omega\sin\alpha\mathbf{i}_T,$$

$$\mathbf{H}_B = J\omega\mathbf{j}_B - J\Omega\cos\alpha\mathbf{j}_B - I\Omega\sin\alpha\mathbf{i}_B, \qquad (8)$$

where $\mathbf{j}_T$, $\mathbf{j}_B$, $\mathbf{i}_T$, and $\mathbf{i}_B$ are unit vectors as in Fig. 11. To find the net angular momentum of the system, one adds $\mathbf{H}_T$ and $\mathbf{H}_B$ and resolves the components of $(\mathbf{j}_T + \mathbf{j}_B)$ and $(\mathbf{i}_T - \mathbf{i}_B)$ along the axis bb'. If the net angular momentum is zero, one finds that

$$\frac{\omega}{\Omega} = \frac{1 + \cos^2\alpha(J/I - 1)}{(J/I)\cos\alpha}. \qquad (9)$$

For example, if $I = J$ and $\alpha = 60°$, then $\omega/\Omega = 2$, and thus if the plumb-bob shaped objects roll once around on their conical ends, the entire body will rotate exactly 180° about the axis bb'.

This demonstrates that it is possible for a body to twist even though at all times during the twist: (i) the net angular momentum of the body is zero; (ii) no external torques are applied to the body; (iii) the principal moments of inertia of the body remain constant; (iv) the "shape" of the body does not change. Of course, twisting is possible only because the body is not truly a rigid body.

In principal a diver or trampolinist with good flexibility in the hip and waist region could perform a twist in a similar fashion. His arms and upper body would be analogous to body "T", in Fig. 11, and his legs would be like body "B." Kane and Scher[9] performed a detailed mechanical analysis on photos of a twisting cat, and concluded that this kind of mechanism was responsible for the twist. McDonald[8] has published drawings of photographs of a twisting man and has suggested that this kind of mechanism may be at least partly responsible for the twist. However, the present author does not agree that most torque-free twists with zero angular momentum are performed in this way by divers, trampolinists, or cats. In practice the performance of these twists is accompanied with considerable motion of the arms (or forelegs) relative to the upper body. In addition, during

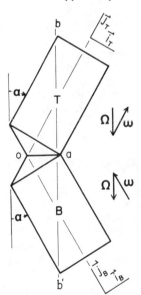

Fig. 11. Diagram illustrating that torque-free twisting can occur even though throughout the twisting process, the angular momentum is zero, the principal moments of inertia of the body are constant, and the shape of the body does not change. The body consists of two identical objects T and B, each shaped like a plumb bob (a cylinder with a cone on one end). If objects T and B each rotate with angular velocity ω about axes $\mathbf{j}_T$ and $\mathbf{j}_B$, respectively, then to conserve angular momentum the entire system must twist with angular velocity Ω about axis bb'. During the process, the concial ends roll without slipping and touch along the line Oa, which rotates in space with angular velocity Ω.

the twist the body generally is bent forward at the waist to a greater degree than to the back or to the sides. Both of these observations favor the hypothesis that the twist is initiated by varying the relative moments of inertia of body parts as in "swivel hips" (Fig. 10).

IV. OBSERVATIONS AND CONCLUSIONS

One of the main conclusions of this paper is that performers can and do execute torque-free somersaults, but only in such a manner that angular momentum is conserved. Uncertainty about this matter exists among divers and trampolinists as well as physicists (see Table I). For example, Rackham[4] says that if a diver jumps from the board with no angular momentum:

No action on the part of the diver during a jump can cause him to rotate to enter the water headfirst. Whatever he does when he finally straightens for his entry he will be in the same position relative to the vertical as when he started the jump.

In fact, without bending his head or trunk, a diver could enter the water head first by twice tucking and untucking his body using the sequence of motion shown in Fig. 6.

As an experimental demonstration of this type of somersault, the author recently performed the following test on a trampoline to show that he could choose to rotate his body 90° after he was in the air. On each designated test jump, a friend watched the author's feet and after his feet had left the trampoline the friend either shouted "Back!" or remained silent. The author then attempted to perform a back one-quarter somersault (see Fig. 6) if "Back!" was called, or a straight jump (no somersault) if nothing was called. To prevent guessing the call, the author's friend determined the calls prior to each test jump by tossing a coin. Of the 20 test jumps that were attempted, the author performed the stunt as called by his friend 19 times.

In terms of physics, the torque-free somersault is closely related to the torque-free twist with zero angular momentum (cat twist). In both stunts motion in one portion of the body possesses angular momentum which must be countered by opposite motion of the remainder of the body. In both stunts rotation is not continuous—it ceases if the performer holds his body rigid. In practice, human performers are generally much more adept at torque-free twists than at torque-free somersaults. For example, Rackham[4] reports that a diver named Brian Phelps could hang by his hands from a ten meter platform, let go, and then do two 180° cat twists—one to the left and one to the right before entering the water.

Incidentally, using both cat twists and torque-free somersaults in combination would allow a hypothetical "spaceman" with zero angular momentum to position his body in any orientation in space that he chooses. This conclusion is in accord with NASA's own research on this question,[20] although the limb movements proposed in the NASA work generally are not used by divers and trampolinists. More often astronauts carry small gas guns which apply torques in order to effect body reorientations.

Another major conclusion of this paper is that when a performer does have angular momentum, "throwing" a twist can indeed produce a continuous rotating motion about his head-to-toe axis even if no torques are applied. Dyson[2] expressed uncertainty about this matter. He noted that

... a gymnast somersaulting forward ... can originate movements in the air which lead to displacement of mass about the main axis. In consequence, his body absorbs some angular momentum by twisting about the longitudinal axis, automatically reducing a tendency to somersault. Conversely, twisting can be "traded" for somersaulting. Thus ∴ .. gymnasts, divers, etc. ... sometimes appear to acquire angular momentum while free in space. However, other reputable students of this subject deny that a "trading" of angular momentum is possible. Here, then, is an important hypothesis requiring more solid experimental evidence.

Other recent authors accept the existence of this kind of twisting.[3,5,12] For example, Batterman[3] presents excellent pictures of a diver showing that his body is inclined to the vertical [as in Fig. 8(b)] while performing a back one and one-half somersault with two and one-half twists. A rather difficult dive which graphically illustrates the possibility of performing torque-free twists with angular momentum is the forward two and one-half somersault with two twists, which is performed from the 3-m springboard.[18] Generally, after leaving the board, the diver executes the first one and one-half somersaults, and only then "throws" for a twist to perform the last somersault with two twists before entering the water.

In practice, do performers really initiate twisting only after they have left the board? Recently the author viewed 64 frame/sec films of divers taken by Richard Gilbert at the 1972 Olympic Trials. For each twisting somersault, the author noted whether the diver dropped his arm or shoulder to initiate twisting before or after his feet left the board. These films revealed significant variations between different divers with respect to when they initiated twisting motion, and with respect to how fast they twisted. Although more than half of the twists were initiated with the feet still on the board for forward somersaults with twists, a significant fraction of the twists were initiated only after the diver was well away from the board. Twisting was more likely to be initiated away from the board for dives with only one or two twists. However, pure torque-free twisting was observed for a dive as complex as a forward one and one-half somersault with three twists.

If a diver initiates twisting after leaving the board, how fast can he twist while executing a forward twisting somersault? Equation (7) in Sec. III reveals that this will depend on the angle θ (see Fig. 9). Although the ratio I_1/I_2 will vary slightly from diver to diver, an individual diver has much more control over the $\sin \theta$ term in (7) than over the ratio of his principal moments of inertia. Exactly how he "throws" will determine how fast he twists.

For example, suppose the diver throws the twist from the layout position (Fig. 8), bringing one arm up over his head and the other arm down to his side. If we think of this as a rotation of his arms 90° relative to his body, Eq. (A6) from the Appendix tells us that if $I_1 = 2.31$ kg m^2 and if the hinge point of the arms is directly between the shoulders $\theta = 10.85°$. Thus from Eq. (7) $\omega_t/\omega_s = 3.0$ twists/somersault.

On the other hand, suppose he "throws" with his arms while he is still in a loose pike position (F in Fig. 1) and then straightens to a layout position to do the twists. Since his body's moment of inertia is smaller in the pike than in the layout (Table III), Eq. (A5) tells us that: $\theta = 19.96°$; and

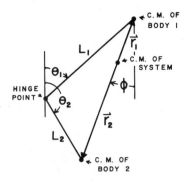

Fig. 12. Variables describing hinged diver. Many body motions of a diver can be interpreted as if the diver's body consisted of two rigid objects connected by a hinge. For example, as the diver pikes or "jackknifes" the hinge point is at his waist, one rigid body is his legs, and the other rigid body consists of his trunk, head and arms.

so $\omega_t/\omega_s = 5.5$ twists/somersault. Thus after he straightens to the layout position he can twist significantly faster because he threw the twist while still in the pike. In practice, motions of the diver's head and trunk as well as his arms will help him to increase the angle θ, and thus his rate of twisting.

The results above are in excellent quantitative accord with the observations of diving films. In the films of the 1972 Olympic Trials, divers performing multiple twisting somersaults were observed to twist at average rates varying from 3.8 twists/somersault to 6.0 twists/somersault. As mentioned previously, one diver performed a forward one and one-half somersault with three twists and clearly initiated the twists well after leaving the board. He was observed to initiate twisting in the pike position, and to complete one-half twist after somersaulting 185°, and two and one-half twists after somersaulting 315°, for an average rate of exactly 5.5 twists/somersault.[21]

ACKNOWLEDGMENTS

Several individuals besides the author made significant contributions to this work. Joe Burns, of the Department of Theoretical and Applied Mechanics at Cornell University, carefully reviewed an earlier draft of this manuscript, as did Charles Batterman, diving coach at the Massachusetts Institute of Technology. I am particularly indebted to Ted Grand, at the Oregon Primate Center, for critical discussions, and to Richard Gilbert, diving coach at Cornell University, for allowing me to view his film collection. Michael E. Fisher, of the Department of Chemistry, Physics, and Mathematics at Cornell, provided the analytical solution to Eq. (A4) in the Appendix after I provoked him by stating that it was not integrable. I also thank the members of the National Collegiate Athletic Association, (NCAA) Diving Rules Subcommittee who took the time to answer the diving questionnaire. Most of all, however, I am indebted to Ray Obermiller, diving coach at Grinnell College, who taught me almost everything I know about diving.

APPENDIX: THE DIVER AS A HINGED BODY

Suppose that a diver behaves mechanically like two rigid bodies which make angles θ_1 and θ_2 with the vertical and are attached to one another by a hinge so that the two bodies are constrained to move in a plane. We wish to calculate how θ_1 and θ_2, change as the diver varies the hinge angle $(\theta_2 - \theta_1)$. Call the masses of the two bodies M_1 and M_2, and the moments about their individual centers of mass

I_1 and I_2, and the distance between the hinge point and their individual centers of mass L_1 and L_2 (Fig. 12). Let ϕ be the angle that the line between the centers of mass makes with the vertical.

If r_1 and r_2 are the position vectors of the centers of mass of the individual bodies in a coordinate system whose origin is at the diver's center of mass, then his angular momentum H of the system is

$$H = I_1\dot{\theta}_1 + M_1(r_1 \times \dot{r}_1) + I_2\dot{\theta}_2 + M_2(r_2 \times \dot{r}_2).$$

This expression can be simplified, since $|r_1 \times \dot{r}_1| = r_1^2\dot{\phi}$. In center of mass coordinates, if $r = r_2 - r_1$, then

$$r_1 = [M_2/(M_1 + M_2)]r$$

and

$$r_2 = [M_1/(M_1 + M_2)]r.$$

Thus the angular momentum H becomes

$$H = I_1\dot{\theta}_1 + M_1 r_1^2\dot{\phi} + I_2\dot{\theta}_2 + M_2 r_2^2\dot{\phi}$$
$$= I_1\dot{\theta}_1 + I_2\dot{\theta}_2 + M_*r^2\dot{\phi}, \tag{A1}$$

where M_* is the reduced mass, i.e.,

$$M_* = M_1 M_2/(M_1 + M_2).$$

To eliminate ϕ and r from this equation, note that

$$\tan\phi = (L_1\sin\theta_1 - L_2\sin\theta_2)/(L_1\cos\theta_1 - L_2\cos\theta_2).$$

Differentiating, we find that

$$r^2\dot{\phi} = [L_1^2 - L_1L_2\cos(\theta_2 - \theta_1)]\dot{\theta}_1 + [L_2^2 - L_1L_2\cos(\theta_2 - \theta_1)]\dot{\theta}_2, \tag{A2}$$

where

$$r = |r_2 - r_1| = [L_1^2 - 2L_1L_2\cos(\theta_2 - \theta_1) + L_2^2]^{1/2}.$$

Substituting (A2) into (A1), we find that

$$H = \{I_1 + M_*L_1[L_1 - L_2\cos(\theta_2 - \theta_1)]\}\dot{\theta}_1 + \{I_2 + M_*L_2[L_2 - L_1\cos(\theta_2 - \theta_1)]\}\dot{\theta}_2. \tag{A3}$$

In particular, if $H = 0$,

$$\dot{\theta}_2 = -\frac{I_1 + M_*L_1[L_1 - L_2\cos(\theta_2 - \theta_1)]}{I_2 + M_*L_2[L_2 - L_1\cos(\theta_2 - \theta_1)]}\dot{\theta}_1. \tag{A4}$$

Surprisingly enough, this expression can be solved exactly in closed form.[22] If neither L_1 nor L_2 are zero, and if we define

$$A_1 = \frac{I_1 + M_*L_1^2}{M_*L_1L_2}; \quad A_2 = \frac{I_2 + M_*L_2^2}{M_*L_1L_2},$$

$$a_+ = \frac{A_1 + A_2}{2}; \quad a_- = \frac{A_1 - A_2}{2}$$

and if $h = (\theta_2 - \theta_1)$; then one can show easily that

$$\frac{d\theta_2}{dh} = \frac{1}{2}\left(1 + \frac{a_-}{a_+ - \cos(h)}\right).$$

This is integrable, and in fact if the hinge angle h changes from h_i to h_f, then the change in θ_2 is

$$\theta_{2f} - \theta_{2i} = \left\{\frac{h}{2} + \frac{a_-}{(a_+^2 - 1)^{1/2}} \times \tan^{-1}\left[\left(\frac{a_+ + 1}{a_- - 1}\right)^{1/2}\tan\left(\frac{h}{2}\right)\right]\right\}\Big|_{h_i}^{h_f}, \tag{A5}$$

where the indefinite integral (in braces) is evaluated at h_f and h_i.

Equation (A4) simplifies considerably if $L_1 = 0$, with the first body having its center of mass at the hinge point. In this case

$$\dot{\theta}_2 = [-I_1/(I_2 + M_*L_2^2)]\,\dot{\theta}_1.$$

If $\Delta\theta_2$ and Δh are the changes in θ_2 and h, respectively, then this becomes

$$\Delta\theta_2 = [-I_1/(I_1 + I_2 + M_*L_2^2)]\Delta h.$$

Since $M_* L_2^2 = M_1 r_1^2 + M_2 r_2^2$ this is just

$$\Delta\theta_2 = (-I_1/I_{\text{tot}})\,\Delta h, \qquad (A6)$$

where I_{tot} is the diver's total moment of inertia about his center of mass.

Equation (A4) is equivalent to Eq. 3.2.7 of Smith and Kane.[10] Smith and Kane[10] also derive expressions to describe more general motions of bodies having two and three segments.

[1] National Collegiate Athletic Association, *The Official Swimming Guide 1978* (National Collegiate Athletic Association, Shawnee Mission, KS, 1977).

[2] G. Dyson, *The Mechanics of Athletics,* 3rd ed. (University of London, London, 1964).

[3] C. Batterman, *The Techniques of Springboard Diving* (MIT, Cambridge, MA, 1968).

[4] G. Rackham, *Diving Complete* (Faber and Faber, London, 1975).

[5] J. Hay, *The Biomechanics of Sports Techniques,* 2nd ed. (Prentice Hall, Englewood Cliffs, NJ, 1978).

[6] Editor, Nature **51**, 80 (1894).

[7] D. McDonald, New Sci. **7**, 1647 (1960).

[8] D. McDonald, New Sci., **10**, 501 (1961).

[9] T. R. Kane and M. P. Scher, Int. J. Solids Struct. **5**, 663 (1969).

[10] P. G. Smith and T. R. Kane, The Reorientation of a Human Being in Free Fall (Technical Report No. 171, Stanford University, U.S. Government Accession Number N67-31537, 1967) (unpublished).

[11] E. P. Hanavan, American Institute of Aeronautics and Astronautics Paper 65-498, 1 (1965).

[12] J. Broms, W. Duquet, and M. Hebbelinck, Med. Sport **8**, Suppl. Biomech. III, 429 (1973).

[13] D. I. Miller, Med. Sport **8**, 116 (1973).

[14] W. T. Dempster, Space Requirements of the Seated Operator (Wright Air Development Center WADC-TR-55-159, Ohio, 1955) (unpublished).

[15] M. Williams, and H. R. Lissner, *Biomechanics of Human Motion* (Sanders, Philadelphia, 1962).

[16] W. R. Santschii, J. DuBois, and C. Omoto, *Moments of Inertia and Centers of Mass of the Living Human Body* (Technical Documentary No. AMRC-TDR-63-36, Wright-patterson Air Force Base, Ohio, 1963) (unpublished).

[17] L. D. Landau and E. M. Lifshitz, *Mechanics* (Pergamon, Oxford, 1965).

[18] *Official Diving Rules 1978,* edited by J. Walker, (Amateur Athletic Union of the U.S., Indianapolis, IN, 1978).

[19] It is possible to obtain a closed solution for the diver's motion without this second assumption. See, e.g., Ref. 17.

[20] T. R. Kane and M. P. Scher, J. Biomech. **3**, 39 (1970).

[21] Perhaps the earliest recorded reference to a twisting somersault appeared in 1674 in *Grammatica Linguae Anglicanae* by John Wallis, a grammar published in Oxford, England:

> When a Twister, a-twisting, will twist him a twist;
> For the twisting of his twist, he three times doth intwist;
> But, if one of the twists of the twist do untwist,
> The twine that untwisteth, untwisteth the twist.
> Untwirling the twine that untwisteth between,
> He twirls, with his twister, the two in a twine;
> Then, twice having twisted the twines of the twine,
> He twisteth, the twine he has twined, in twain.
> The twain that, in twining, before in the twine;
> As twins were untwisted, he now doth untwine;
> Twixt the twain intertwisting a twine more between,
> He, twirling his twister, makes a twist of the twine.

[22] Michael E. Fischer (private communication).

Physics of the tennis racket

H. Brody
Physics Department, University of Pennsylvania, Philadelphia, Pennsylvania 19104
(Received 23 October 1978; accepted 8 February 1979)

Several parameters concerning the performance of tennis rackets are examined both theoretically and experimentally. Information is obtained about the location of the center of percussion, the time a ball spends in contact with the strings, the period of oscillation of a tennis racket, and the coefficient of restitution of a tennis ball. From these data it may be possible to design a racket with improved playing characteristics.

Reprinted from *American Journal of Physics*, **47**, 6. ©1979 American Association of Physics Teachers.

The physics of the tennis racket is a subject that very few people paid attention to until Head produced his over-sized Prince racket. In fact, in reading various tennis magazines, looking at advertisements for tennis rackets, and talking to tennis players, tennis professionals, or sports equipment sales people one gets such conflicting statements that one realizes that no one seems to understand the physics of the tennis racket.

When a tennis racket is examined, one notes that the basic shape and size have not changed in over half a century. It is probable that these parameters were not determined solely by playing characteristics but also by structural considerations imposed by the strength of wood. When metal rackets became popular a few years ago (and composite fiberglass, boron, or carbon filament recently) the manufacturers initially copied the general shape and size of the wooden rackets, since they probably assumed that they were optimum—or that any radical change might not sell. The same might be said for the strings—where gut has successfully withstood the challenge of all the modern synthetic materials for tournament play.

There seems to be no information in the published physics or tennis literature about the optimization of size, shape, weight, etc., of a tennis racket. Since everyone learned to play with essentially the same type of racket, any radical change would feel wrong and require the player to relearn to some degree. Under these conditions the design of rackets might evolve and improve slowly—but there is no way to know if a racket that was radically different might not prove better if a player originally learned with it.

One obvious change was the introduction of the Prince racket with an oversized head. Head, its inventor, was very careful to produce a racket of the same overall length, weight, and balance as a conventional racket so that it would "feel" the same as a normal racket. To prove this, any average player can pick one up and play with it immediately. However to play with it in an optimum manner does take some retraining since it was designed to give its best performance with the ball striking it 5 or 6 cm closer to the handle than on a normal racket. To understand the advantages of the Prince racket (without playing with it) some simple kinematics must be investigated.

Consider a tennis racket of mass M suspended freely in space. If a ball strikes it and imparts a momentum $+ \Delta p$ to the racket, the center of mass of the racket will move with a velocity $V = + \Delta p/M$.

If the ball hits the racket at the racket center of mass, the racket will translate but not rotate. If the ball hits the racket

at a distance b from the center of mass, the racket will rotate around the center of mass as well as translate. From conservation of angular momentum, the angular velocity of the racket about the c.m. will be $\omega = \Delta pb/I_{c.m.}$ where $I_{c.m.}$ is the moment of inertia of the racket about its center of mass. Assume the racket has been struck by a ball, the racket (c.m.) is translating to the right with velocity V and the racket is rotating clockwise about the c.m. with angular velocity ω [Fig. 1(a)]. There then will be one point in the racket which is instantaneously at rest, if the racket handle is long enough. If this point is a distance a from the center of mass then the condition for that point to have no velocity is $V = \omega a$. In other words, the motion due to the rotation exactly cancels the overall translation (of the racket) at that one point. Then

$$ a = \frac{V}{\omega} = \frac{\Delta p/M}{\Delta pb/I_{c.m.}} = \frac{I_{c.m.}}{b\,M}. $$

The value of a is independent of how hard the ball hits the racket (Δp) as it should be. Then $ab = I_{c.m.}/M = k^2$ (radius of gyration squared).

If a ball hits a racket at a distance b from the c.m. (at point B) and the racket is held a distance a from the c.m. (at point A), no force or impulse from the hand need be imparted to the racket, since, in a frame of reference initially moving with the racket, the point A remains at rest. In the frame of reference with the racket handle being swung with velocity V', the point A continues to move with velocity V' with no external force applied to it.

If the racket is held at point A then point B is called the center of percussion and it is of some interest to determine the distance $a + b$. This distance $a + b$ can be obtained with a simple experiment that uses the racket as a physical pendulum. If the racket is allowed to swing freely about the point B, the frequency of the oscillation (for small amplitude) can be calculated and also measured.

In Fig. 1(b), the restoring torque is $-Mg b \sin \theta \sim -Mg b \theta$ (for small θ). Then $\tau = I_B (d^2\theta/dt^2)$, where I_B is the moment of inertia about B. Using the parallel axis theorem, $I_B = I_{cm} + Mb^2$ and substituting this into the $\tau = I\alpha$ equation

$$ -Mg b\, \theta = (I_{cm} + Mb^2) \frac{d^2\theta}{dt^2}. $$

The solution to this equation is a simple harmonic motion with angular frequency $\omega = [Mg b/(I_{cm} + Mb^2)]^{1/2}$ but $I_{c.m.} = Mk^2$, so $\omega = [gb/(k^2 + b^2)]^{1/2}$. However, since $k^2 = ab$, $\omega = [gb/(ab + b^2)]^{1/2} = [g/(a + b)]^{1/2}$. Conse-

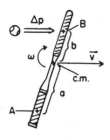

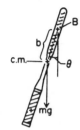

Fig. 1. (a) Tennis racket suspended freely in space; (b) Tennis racket as a physical pendulum with pivot at center of head.

quently, from a measurement of the angular frequency of the racket used as a pendulum and a knowledge of g, $a + b = g/\omega^2$ can be found.

In actually doing the measurement, care must be exercised to make the plane of swing perpendicular and not parallel to the face of the racket since the moments of inertia are different. For a normal racket held at the handle, the center of percussion is 5 or 6 cm below the center of the head of the racket. (Since these equations are symmetric in a and b, $a + b$ could also be found by allowing the racket to pivot about the point where the racket is held.) Three rackets (Spalding Smasher, Davis TAD, and Prince) were suspended with a pivot point at the center of their head and allowed to oscillate as a pendulum. Their period was measured and the distance $(a + b = g/\omega^2)$ calculated. The distance of the conjugate center of percussion point from the butt end of the racket was then obtained. These data are shown in Table I. If a racket was held at the conjugate point, then the center of percussion would be at the center of the head of the racket. If you desire to hit a ball at the center of percussion of a normal racket you should either "choke up" by moving your hand up the handle 5 or 6 cm and hit the ball at the center of the head or keep your hand in its normal position and hit the ball close to the throat of the racket. Neither of these is recommended.

A third choice is to build a different racket so that when the racket is held in the normal place, the center of percussion is at the center of the stringed region of the head. This can be done by making the tip of the racket very massive or by extending the head of the racket considerably in the direction of the handle. It is this latter technique that Head employed in his Prince racket.

There is, of course, the question as to whether the center of percussion is the optimum location to hit a ball. If the ball spends an appreciable amount of time in contact with the strings of the racket (a long dwell time) then it might be possible for a player to add momentum and energy to the racket, hence the ball, during the dwell time. However, because of the conjugate nature of the two points, the motion of the center of percussion will not be affected by a simple force applied at the handle, hence it may not be possible to affect the ball during its dwell. A torque about the center of the handle will, however, affect the ball. Since

the dwell time of a tennis ball on the strings of a racket is about 5 msec, the idea of hitting the ball at the center of percussion of the racket bears further examination but not in this paper.

If a ball is hit at the center of percussion there will be no net force on the hand. However, in the hand frame of reference the racket will attempt to rotate about an axis perpendicular to the handle at the point where it is held. If the ball is hit further away from the handle than the center of percussion, as it is in most rackets, then the racket (in the hand frame of reference) will attempt to rotate about an axis passing through the handle at a point closer to the head of the racket. On forehand there will then be a net force on the hand (due to the ball interaction) that will tend to pull the racket forward out of the player's hand (tending to force open the fingers and thumb).

If the ball is hit closer to the handle than the center of percussion (which is almost impossible on a normal racket) then the racket will attempt to rotate about a point near the butt end of the handle. There will again be a net force or impulse on the hand, only now (for forehand) it will be directed against the palm and base of the fingers. From this analysis it would seem advantageous to hit the ball closer to the handle rather than farther away, and this is reinforced by the following analysis. If the racket is treated as a free body (not held in a hand) it is clear from the two body kinematics that the closer to the center of mass of the racket that the ball strikes, the higher will be the ball's rebound velocity. This is because less energy goes into the rotation of the racket when the ball hits closer to the c.m. Consequently, enlarging the head by extending it toward the handle has definite advantages. This would lead to a very elongated shape unless the head were made wider, and there is a good argument for doing this as well.

Many players, particularly beginners, have a problem that the racket twists or rotates in their hand. This is because the ball has hit off center (not along the polar axis) and a net angular impulse is imparted to the racket. The further from the axis the ball hits, the more the racket will tend to twist. This twisting can be reduced by increasing the polar moment of inertia of the racket. This can be accomplished by increasing the mass along the sides of the head or by keeping the mass fixed and increasing the distance from axis to the side of the head (in other words, making the head larger).

The Wilson company has chosen the former by adding tungsten weights in a perimeter weighted racket. Head has opted for the latter in his Prince racket. Since the moment of inertia goes as mR^2, an increase in size of 40% results in

Table I. Racket parameters.

Racket	Davis TAD	Spalding Smasher	Prince
Length (cm)	68.9	67.0	68.6
Period of oscillation about center of head (sec)	1.26	1.28	1.30
$a + b = g/\omega^2$ (cm) distance between pivot and center of percussion	39.4	40.7	42.0
Distance between center of head and butt end (cm)	54.6	54.0	51.1
Distance between center of percussion conjugate point and butt end (cm)	15.2	13.3	9.1

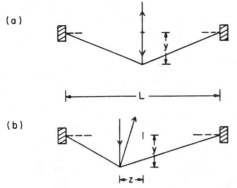

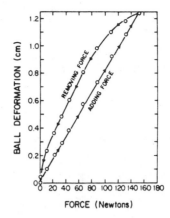

Fig. 4. Deformation of a tennis ball as a function of the applied force. The ball was placed in a rigid hemispherical cup so that only one side deformed.

Fig. 2. String deflection when ball hits (a) at center of strings and (b) closer to one side of racket.

a doubling of the moment of inertia. It is also possible to increase the diameter of the racket handle so that the torque tending to prevent twisting is increased without increasing the forces applied by hand.

Even if the ball is hit off center and the racket twists a little, there is another effect that will tend to compensate for it. If the ball hits off center, the deflection of the strings is asymmetric. Neglecting spin of the ball, gravity, etc., in the rest frame of the racket, it is expected that relative to the plane of the strings the angle of the ball's rebound will be equal to the angle of its incidence. This will be true only if the strings deflect symmetrically [Fig. 2(a)] which occurs for hits in the center. For off center hits, the asymmetrical string deflection tends to deflect the ball toward the center [Fig. 2(b)]. Head was able to demonstrate this using a high-speed motion picture camera with the racket held in a vise. The data he obtained allowed him to determine the angular error of the rebounding ball as a function of position and map out contours of angular error regions with error less than 10°, between 10° and 20° and greater than 20°.[1]

This angular error depends upon several factors. If a string of length L deflects (perpendicular to the plane of the strings) by a distance y when hit by the ball and the ball misses the center by a distance z, the angular error will be proportional to $(z/L)(y/L)$. There is not much a physicist can say about reducing z, hence reducing the error, but y and L are subject to analysis. The maximum value of the string deflection can be obtained if the effective spring constant of the strings is known (the slope in Fig. 5), the ball momentum specified and the assumption that the motion

is simple harmonic is made. Then $\int F\,dt$ over $\frac{1}{2}$ cycle equals Δmv. Since $F = -ky$ and $y = A \sin \omega t$, the amplitude A of the deflection can be obtained: $A = (\Delta v/2) (m/k)^{1/2}$, where m is the mass of the ball. (The mass of the strings is neglected.) The value of k is a linear function of T/L, the string tension divided by the string length and also depends upon the effective number of strings that deflect. When all of this is put together, the angular error $\Delta\theta = (Cz\,\Delta v/L)(m/LT)^{1/2}$, where C is a constant.

It is then quite clear that increasing the size of the racket head (increasing L) reduces this error, hence increases the effective area of the racket within which this error is tolerable. It is also clear that an increase in string tension will reduce this error somewhat.

Another interesting result is that this error is proportional to Δv, the change in the ball velocity. Consequently, when the ball is hit hard, unless it is hit dead center on the racket, it may not go exactly where it is aimed. This compounds the difficulty of hard hits, which, due to their kinematics, already have very little margin for error.

The string tension also influences how long a ball spends in contact with the strings (dwell time) and the velocity at which the ball leaves the racket.[2] To optimize these parameters various measurements have to be made, including a determination of some of the properties of tennis balls.

The most surprising thing is that it appears that stringing the racket looser rather than tighter will actually lead to slightly higher ball rebound velocities (more "power"). This is due to the fact that tennis balls have a rather low coefficient of restitution and a dwell time on the strings which is short when compared to half of the natural period of vi-

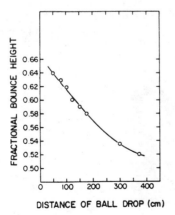

Fig. 3. Ratio of rebound height to drop distance as a function of drop distance. Ball was dropped onto a hard surface.

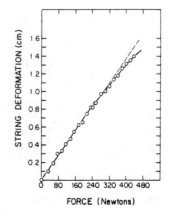

Fig. 5. String deformation as a function of applied force. The racket head was braced and the force was applied over a circular area of 12 cm².

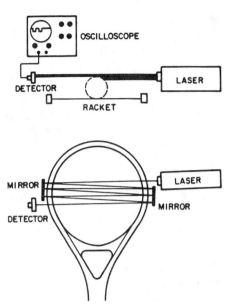

Fig. 6. Method for measuring the dwell time of a tennis ball on the strings of a racket. The head of the racket is firmly clamped in place and the laser beam is parallel to and one ball diameter above the plane of the strings.

bration of many tennis rackets. To see how this all fits together, let us examine some simple tests on balls and rackets that were made.

When a tennis ball is dropped from a specified height onto a hard surface, the height of the ball's rebound is determined by its coefficient of restitution (COR). The ratio of final height to initial height is (COR)^2 if air resistance is neglected. The COR of several tennis balls was measured by this method and also measured directly by determining the ball velocity using electronic techniques. It was discovered that the higher the ball velocity upon impact, the lower the COR for that collision, and that about half of the kinetic energy of a ball is lost when it hits a hard surface. A measurement of ball deformation versus applied force shows a large hysteresis loop, which is indicative of a large loss of energy. The strings (and the racket frame) have relatively little loss of energy when they are deformed. If the strings can take up a larger fraction of the energy, the ball will deform less and lose less energy. Consequently the ball will rebound with more energy (fed to it by the strings) and will have a higher velocity. The less the tension in the strings, the more they deform the larger the fraction of energy they store (and subsequently give back), the less the ball deforms, the less kinetic energy the ball dissipates. There is a limit to how loose the strings should be strung, since a butterfly net is clearly not any good. Once the tension is reduced to a point where the strings being to move in a direction parallel to the plane of the racket head, the energy that goes into that mode is probably not recovered and the strings will probably wear out because they are rubbing against each other.

The above analysis will only hold if the strings are able to feed the energy they absorb (or store) back into the ball. This means that the ball cannot come off of the strings before the strings come back to their undeformed position. This could only happen if the time required for the deformed ball to revert back to its original spherical shape is short compared to the time required for the strings to snap back. Electronic measurement of the time the ball spends

in contact with a hard surface and the strings on a racket show that this is not the case.

There is a much easier way to experimentally demonstrate the above concept. If a tennis ball is dropped onto a hard surface from a height of 100 cm, it rebounds to a height of about 62–65 cm. If the same ball is dropped from a height of 100 cm above the head of a tennis racket (resting on the floor) it will rebound to a height of slightly over 80 cm. Measurement of the ratio of rebound velocity to the incident velocity, obtained using electronic means are in agreement with the rebound height measurements.

One would naively assume that a similar argument would hold with respect to a stiff versus a flexible racket frame. By the above arguments, a more flexible racket would lead to more energy being returned to the ball. However, due to its mass, the natural half period of vibration of a racket can be long compared to the dwell time of the ball on the strings. Consequently, most of the energy that goes into deflecting or deforming the racket is *not* fed back into the ball. The ball is gone by the time the racket snaps back. Therefore a stiffer racket would provide more power since less energy would go into the racket frame deformation and also the natural period of oscillation is reduced. If the racket oscillation time can be reduced so that half a period is comparable to the ball dwell time, the energy from the racket deformation will be fed back into the ball's kinetic energy.

When the tension in the strings is reduced, the dwell time of the ball is increased, and this can also improve the match between dwell time and racket oscillation period.

The great loss of energy to the racket deformation can be seen by bouncing a ball off of the racket with the handle in a vise measuring the COR as a function of the position on the head where the ball hits. The COR varies from 0.6 very close to the throat of the racket to 0.2 near the tip of the racket with values between 0.4 and 0.5 at the center of the head. This is because the stiffness of the racket changes with position, with little flex close to the throat and much deflection near the tip. Consequently more energy goes into racket deformation when the ball is struck near the tip.

If the COR tests are done with the head of the racket clamped (instead of the handle) the COR is much higher and it maximizes at the center of the head. This is because that is where the strings are effectively least tightly strung and they deflect most. Moving away from center, the strings deflect less, the ball deforms more and more energy is lost.

In actual play, several other factors enter that do not show up in a bench test. As was stated before, if the ball struck closer to the neck, then it is closer to the racket center of mass and less energy goes into racket rotation, hence more energy goes into the ball. However, a racket is swung in an arc with a point in the body as the pivot, so the velocity

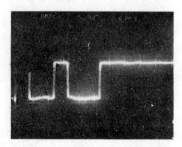

Fig. 7. Picture of an oscilloscope trace measurement of dwell time. Sweep speed is 5 msec per division.

of a particular point on the racket is a function of its distance from that point. Moving the impact point several inches toward the body should reduce the velocity of the racket at the impact point by less than 10%, so this effect can probably be ignored. (It may be somewhat more significant in the serve when the wrist is snapped—which makes the pivot to impact distance considerably shorter.)

The length of time that a ball spends in contact with the strings of the racket can be determined by direct measurement or it can be approximately calculated from static measurements made on the ball and strings. If the string-ball system is treated as a simple harmonic oscillator, then the dwell time of the ball on the strings will be half the natural oscillation period of the system. The effect "k" (spring constant) of the strings is a function of the stringing tension, gauge and type of string, size of the racket head, etc. This k was determined for the strings of several different rackets by measuring the deflection of the strings as a function of applied force. Typical measured values ranged from 2 to 3.5×10^4 N/m (the slope of Fig. 5). A tennis ball has a mass of 0.060 kg, leading to a natural frequency of oscillation of 100h [$f = \omega/2\pi = (1/2\pi)(k/M)^{1/2}$] and a dwell time of 4.5×10^{-3} sec.

The direct measurement of dwell time can be accomplished in a number of ways, i.e., by using high-speed motion picture photography (greater than 1000 frames/sec), by taking a series of single flash pictures of a repetitive event—each picture at a slightly different electronic delay or by a method using a laser, photodetector, and an oscilloscope that I have developed.

The apparatus for the laser method is shown in Fig. 6. The ball is placed on the racket strings and the laser is adjusted in position so that its beam is parallel to the rebounding surface and is exactly one ball diameter above it (part of the beam is intercepted by the ball at rest). A photodetector (silicon solar cell) is positioned in the beam beyond the ball and the output of the detector fed to an oscilloscope on internal trigger. The ball is then removed.

The ball is then dropped onto or propelled at the rebounding surface. The ball passing through the laser beam blocks the light from the detector and triggers the scope sweep. The light will again hit the detector when the ball has passed completely through the beam. However, due to the position of the laser and rebounding surface, this will only occur while the ball is in contact with the surface (strings). As the ball leaves the surface, it again intercepts the laser beam and cuts off the light to the detector. The resulting scope trace shows a period of time with no light (ball moving toward surface), a period of time with light (ball in contact) and a second period of no light (ball rebounding). This method not only gives the dwell time, but, since the ball diameter is known, it gives the velocity of incidence and velocity after rebounding—hence the coefficient of restitution. If the ball does not cut through the laser beam at a diameter (the center of the ball misses the beam by a distance Δx) then there will be an error in the timing due to the fact that the beam will be cut by a chord rather than a diameter. The chord, being shorter than the diameter by a distance $4(\Delta x)^2/D$ will cutoff the light for less time and will also increase the apparent dwell time correspondingly. This timing error will be $4(\Delta x)^2/vD$ where v is the velocity of the ball and D its diameter. The fractional error in the time the light is cut off is then $4(\Delta x)^2/D$ and the fractional error in the dwell time (t) will be $4(\Delta x)^2/Dvt$.

Table II. Coefficients of restitution.[a]

Surface	Ratio of rebound height to drop height	COR	COR from velocity measurements
Lead brick	0.520	0.721	0.767
Prince racket (70-lb tension)	0.716	0.846	0.887
Prince racket (50-lb tension)	0.730	0.854	0.901
Spalding Smasher (unknown tension)	0.726	0.852	0.895

[a]These data were taken with the ball dropped from a height of 3.7 m above the rebounding surface. It is the average of data taken with several types of tennis balls (pressureless Tretorn, Penn ball, and Spalding Australian ball). The COR values obtained by direct velocity measurements (laser method) are consistently higher than the rebound measurements by about 0.05. This is due to the air resistance which reduces the kinetic energy of the ball during both its fall and rise, and therefore give lower values for the COR. For these data the racket head was clamped.

For Δx of order 1 cm, $D = 13.2$ cm, $v = 800$ cm/sec and $t = 5 \times 10^{-3}$ sec the time errors are 2% and 7.6%, respectively. To reduce this error, a pair of plane mirrors was set up and the laser beam multiply reflected before hitting the detector. The spacing between adjacent beams was of order 1 cm which leads to a maximum spatial error $\Delta x = 0.5$ cm and a maximum error in dwell time of less than 2%.

A second error in dwell time is present if the laser beam is not exactly one ball diameter above the rebounding surface. If the placement error is Δy, then the dwell time will be in error by $2\Delta y/v$. An error Δy of 2 mm then leads to an error in timing of 0.5×10^{-3} sec, so an effort was made to have the beam within 1 mm of the ball top. The results of these measurements with several different rebounding surfaces are shown in Table III.

The predicted dwell time of 5×10^{-3} sec seems to be confirmed by these measurements. The tennis rackets used were those that were readily available. It is clear that these measurements should be repeated using a number of identical rackets, each strung with different material and a variety of tensions.

The tennis racket frame also has a natural period of vibration which is determined by the mass and mass distribution of the frame, the elasticity or stiffness and the damping of the oscillations. The parameter of interest is

Table III. Typical dwell times (msec).[a]

| Rebounding surface | Type of ball | | |
	Penn	Spalding Australian	Tretorn (pressureless)
Lead brick	3.9	4.5	4.5
Prince (70 lb)	6.1	6.3	6.3
Prince (50 lb)	6.5	6.7	6.4
Spalding Smasher	6.4	6.8	6.6

[a]All balls were dropped from a height of 3.7 m. Their measured velocity at impact (8.3 m/sec) was slightly lower than the theoretical value (8.5 m/sec) because of air resistance. One Prince racket was strung with a synthetic at 70-lb tension, the other Price with gut at 50-lb tension and the Spalding Smasher with synthetic at an unknown tension. The racket head was clamped for these data.

Table IV. Period of oscillation of tennis rackets.[a]

Type of racket	Frequency of oscillation (h)	Time for half-period (msec)
Spalding Smasher	32.8	15.3
Wilson T-3000	25.6	19.6
Prince	31.3	16

[a]These data were taken with the handle clamped firmly to a massive table and the racket struck at the center of the strung area by a ball.

actually a half period ($T/2$) which is the time it takes for the racket to go from its undeflected configuration to fully deflected by the ball and then return to its undeflected or equilibrium position.

The period of oscillation of several tennis rackets was measured by placing the racket handle in a vise with the head free to vibrate. A small mirror (mass $\ll$ 1 g) was taped to the tip of the head. A laser was positioned so that its beam reflected off of the mirror and onto a photodetector with the racket at rest. The output leads from the cell were connected to a scope, internally triggered and set for single sweep at either 10 or 20 msec/division. The racket was then deflected, the trace photographed, and the period of oscillation measured.

The half period of the several rackets measured came out to be approximately twice the dwell time of the ball on the strings. If this is the mode of oscillation that a racket undergoes when it strikes a ball, then all the energy that goes into deforming the racket is lost. This is because the ball has left the strings before the racket can begin to snap back. This energy loss can be eliminated by making a much stiffer racket—one with a natural period of oscillation half as long as a normal racket, stringing the strings at much lower tension, using thinner or more elastic strings, or some combination of all of these. The result will be a racket that will propel the ball with a higher velocity for the same swing by the player.

If a hand held racket tends to oscillate with a node at the center of mass and not at the handle, then the oscillation period is much shorter and the energy of deformation might be fed back into the ball. Unfortunately my measurements do not distinguish between these two cases since they were done with the racket firmly clamped and not hand held.

The tests and calculations show that a racket with a larger head seems to have definitive advantages. It is not clear whether the Prince racket is the optimum size, since an even larger racket may eventually prove to be better. However, if a racket of conventional length, weight and balance is desired, the Prince is an improvement over standard rackets. One possible modification would have been to produce a racket with an oversized head where the center of the strings coincided with the center of percussion and they were located at the same distance from the butt end as the center of the head is in a conventional racket. This would lead to a racket with somewhat different weight and

balance, but retain the hitting distance that players have learned to play with.

ACKNOWLEDGMENTS

I would like to thank Jim Hobbs and Michael Procario for their help in taking some of the data presented in this paper.

APPENDIX

The calculation of dwell time assumes a simple harmonic motion with the restoring force of the strings on the ball proportional to their deflection out of the plane of the racket. This is a good approximation for small deflections (Fig. 5), but is the deflection small when a ball is actually hit? If there is an appreciable deflection, then the cubic term in the force-deflection relationship becomes important and the strings tend to act as if they are stiffer (the effective k increases). The maximum deflection of the strings can be calculated assuming a linear relationship, but knowing that the actual displacement from equilibrium will be less:

$$\Delta mv = \int F dt = \int -kx dt,$$

where x is the displacement from equilibrium. If $x = A \sin \omega t$ where $\omega = (k/m)^{1/2}$, the integral becomes $\Delta mv = \int -kA \sin \omega t dt = -kA/\omega \cos \omega t$. Evaluating this over a full half cycle $\Delta mv = 2kA/\omega = -2kA (m/k)^{1/2}$. Then $A = (\Delta v/2) (m/k)^{1/2}$. Putting in the measured values for the k of the strings (from Fig. 5), $k = 2 \times 10^4$ N/m, the measured mass of a tennis ball, $m = 0.060$ kg and a $\Delta v = 40$ m/sec yields a maximum deflection $A = 3.5$ cm. The $\Delta v = 40$ m/sec corresponds to a well hit ground stroke with the ball approaching at 40 miles/h and leaving at 50 miles/h.

As is evident from Fig. 5 a deflection of 3.5 cm is well into the region where the cubic term becomes important. This means that the ball will spend less time in contact with the strings than the linear calculation would predict. The agreement between the calculation and the dwell time measurements that were made is because the measurements were made with ball velocities that corresponded to $1-1\frac{1}{2}$ cm deflections of the strings where the linear approximation is good.

In the limit of infinite tension in the strings, the dwell time is determined by the ball only. Measurements made under these conditions (bouncing the ball off a brick) gave dwell times of about 4 msec.

[1]Howard Head, U.S. Patent No. 3999756, Dec. 28, 1976.

[2]It is interesting to note that string tension is the one parameter that the average player can easily vary a great deal since a racket can be strung with anywhere from 35 to 75 lb of tension. The other parameters (racket weight, balance, handle circumference, head size, flex, or whip) have a much narrower range of variations—among the readily available commercial rackets. Yet the string tension is probably selected by many players in a completely irrational manner—"Gee, that sounds good."

Physics of the tennis racket II: The "sweet spot"

H. Brody

Physics Department, University of Pennsylvania, Philadelphia, Pennsylvania 19104
(Received 22 September 1980; accepted 17 November 1980)

The term sweet spot is used in describing that point or region of a tennis racket where the ball should be hit for optimum results. There are several definitions of this term, each one based on different physical phenomenon. In this paper the different definitions are discussed and methods are described to locate the points corresponding to each one.

Reprinted from *American Journal of Physics*, **49**, **9**. ©1981 American Association of Physics Teachers.

One of the problems in trying to analyze the "sweet spot" of a tennis racket is that, not only are there several different definitions of sweet spot, but some people consider it a point and others claim that it is a region on the face of the racket. To add to the confusion, most people who use the term, particularly in advertisements, never bother to define it at all.

Howard Head, in his ads for the Prince racket and in his patent,[1] defines the sweet spot to be that region on the racket face where the coefficient of resitution (COR) is above some arbitrary value when the racket handle is firmly clamped. To reduce confusion, this area should be called the power region and the place where the COR maximizes should be called the power point.

A second definition claims that the sweet spot is the center of percussion. This is a well-defined point (or pair of conjugate points) for a rigid body and the location of this point can be found by using the racket as a physical pendulum and measuring its period. Since the center of percussion was covered in the previous paper,[2] it will not be discussed further here.

A third definition has the sweet spot as the point where, when the ball hits, the resulting higher frequency vibrations or oscillations are a minimum. This point is the node and the region around the node where the vibration is below some arbitrary value will be designated as the nodal region.

Many people use the term sweet spot to designate that point or area on the racket where it feels good when you hit the ball. This is quite subjective and hard to define, but it is probably closely related to some combination of the previous definitions.

POWER REGION

There is usually one area on the stringed portion of the face of a racket where the coefficient of restitution (COR) is a maximum. Contours of equal COR can be determined experimentally by having a tennis ball strike the racket at various places and measure the ratio of rebound velocity to incident velocity with the racket handle firmly clamped. Another way to accomplish this is to drop a ball on the face of the racket and measure the ratio of rebound height h_r to drop height h_d The COR is $\sqrt{h_r/h_d}$. It is found that the COR tends to maximize along the longitudinal axis of the racket and peaks fairly close to the throat—not near the center of the head. If the ball, string, and racket system is analyzed, it is quite clear why this is the case. It was shown in paper I that any energy that goes into the racket's deformation is lost since the ball has left the strings by the time the racket springs back. It was also shown that the higher the string tension the more the ball deforms upon impact hence the more energy the ball dissipates. Therefore, to increase the COR, more energy should be absorbed by the strings (lower tension) and less energy go into racket deformation (stiffer frame). Using this simple model, the data for Fig. 1 (COR versus racket stiffness) was generated.

The reason that the maximum COR is along the longitudinal axis and displaced several centimeters up from the throat end of the face is due to the strings being "softest" at the center of the head and the frame being stiffest near the clamped handle. For the strings only (racket head clamped) the power spot would be in the center of the head. This last fact can be demonstrated experimentally by applying a known force to the strings at various locations and measuring the deflection or by calculating the deflection y as a function of positions for a given external force F. Referring to Fig. 2, with a single string of length L and tension T, the restoring force $F = T_1\sin\theta_1 + T_2\sin\theta_2$. Setting $T_1 = T_2 = T$ and $\sin\theta = \tan\theta$ gives

$$F = \frac{Ty}{L/2 - z} + \frac{Ty}{L/2 + z} = Ty\left(\frac{1}{L/2 - z} + \frac{1}{L/2 + z}\right)$$

$$= \frac{TyL}{L^2/4 - z^2}.$$

If $z \ll L$, this can be expanded giving

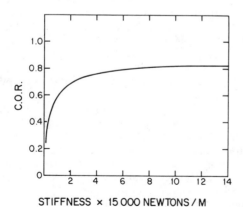

$$\begin{array}{c} 1.0 \\ 0.8 \\ 0.6 \\ 0.4 \\ 0.2 \\ 0 \end{array}$$

STIFFNESS × 15 000 NEWTONS / M

Fig. 1. Coefficient of restitution verses racket stiffness for a ball hitting at the center of the string area when the racket handle is firmly clamped. A typical racket will deflect 1 cm for a force of 100 N, hence will have a stiffness of 10 000 N/m.

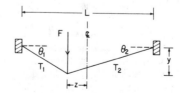

Fig. 2. Deflection y of a string of length L due to a force F applied at a distance Z from its center.

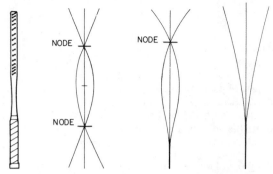

Fig. 3. Vibrations of a tennis racket for the free-free and the clamped-free modes.

$$F = (4Ty/L) (1 + 4z^2/L^2 + \cdots).$$

For a given deflection y, the force will be a minimum at the head center for two separate reasons. The strings that meet at the center of the head are the longest ones in the racket and this reduced the $4Ty/L$ term. (It is assumed that T, the tension, is the same for all strings.) At the center of the head $z = 0$, so the $4z^2/L^2$ term is zero. This would produce a maximum of the COR in the center of the head if the frame were absolutely rigid (or the head clamped). Since it is the handle that is clamped in the COR tests, the frame flexibility must be taken into account. The racket frame deflection of several rackets was measured for a given force as a function of longitudinal position x from the clamp. It varied for different rackets, but the deflection was of the form kx^m where m was of order 2. The racket becomes much stiffer closer to the clamped end. This frame flexibility term dominates over the previously determined string term, shifting maximum COR close to the throat of the racket.

There are a number of ways to increase the COR of a racket. When the head of a racket is enlarged by extending it toward the handle (without increasing the overall racket length) a region of higher COR is produced compared to a racket of standard head size with the same flexibility. To make use of this region of higher COR in actual play, the ball must be hit closer to the handle of the racket. This will lead to a higher ball velocity if the racket motion is translational and all points on the head have about the same velocity. If the racket is rotating (such as in a serve where it is whipped or snapped), the higher COR region, being closer to the effective axis of rotation, is moving with a lower velocity than the region near the racket tip. On this type of swing, maximum ball velocity is not obtained by hitting the ball at the point in the racket where the COR maximizes, but at a point where the racket velocity is high if the racket is not too flexible. It is interesting to note that some people, upon examining their racket, will find two separate areas of wear on the strings. One area is above the center of the head and indicates where they have been hitting the serve. The second area of wear is below the center and is the result of ground strokes.

A second method to increase COR is to make the frame itself stiffer by increasing its cross-sectional area or by going to a new material. When wood was the only material used in racket construction, a stiffer racket meant a heavier racket. With the availability of composite materials, it is now possible to have a racket that is both light and quite stiff while being capable of standing up to the pounding that a racket is subject to.

A third way to increase a racket's COR is to use strings at reduced tension, however, that is not the only important parameter. In examining the formula for string deflection $F = 4T/L\, y\, (1 + 4z^2/L^2 + \cdots)$ it is clear that what is important is the string tension divided by string length and not the tension alone. It is then obvious why, to get a similar "feel," a larger headed racket is strung at a higher tension.

NODE

If the handle end of the racket is firmly clamped, and the racket is struck, it will oscillate. The fundamental mode of vibration has no nodes and a frequency of 25–40 Hz depending upon the mass, mass distribution, and stiffness of the racket (Fig. 3). The next higher mode of oscillation has a single node and, if the racket were a uniform beam of length L, the frequency would be about six times higher than the fundamental and it would have a node a distance $L/5$ from the tip. If the racket were to be considered a free uniform beam of length L (no clamping) its fundamental frequency would also be about six times the clamped fundamental and it would have two nodes, $L/5$ from each end. Many people consider the excitations of this higher mode of oscillation to be undesirable—leading to loss of control, fatigue by the player, and a generally unsatisfactory feeling when the ball is hit. The node is a clearly defined point (or line) while the size of the nodal region is subjective since it is determined by how much amplitude of the higher mode of oscillation one is willing to accept. The next higher modes above these for both the free-free and free-clamped beam are sufficiently high in frequency and damp out so quickly that they can be ignored.

Since a racket is not a uniform beam, but a relatively complicated structure, the frequency and amplitude of the oscillations for the free-free racket and free-clamped racket must be determined experimentally. This has been done in the case of the free-free racket by Lacoste[3] and Hedrick et al.[4] using similar techniques. Hedrick freely suspended a racket, mounted an accelerometer on the handle, and hit the strings at various locations. The accelerometer recorded an impulse plus an oscillation. The impulse is not present when the ball hits the center of percussion (COP) and the oscillation is not present when the ball hits the position of the node of the fundamental free-free mode of oscillation. This latter point is then called the sweet spot by those authors and they have determined its position for a large number of rackets. They did not report the distance between the COP and the node.

To measure the oscillations of a racket with the handle clamped, a method was developed to determine the position of the tip of the racket as a function of time. This method is illustrated in Fig. 4 and a series of photographs was obtained corresponding to a ball hitting at different locations along the polar (longitudinal) axis of a racket. When the ball struck at the location of the node of the higher mode of oscillation, the tip executed a pure simple harmonic motion at the (clamped-free) fundamental frequency [Fig. 5(b)]. As the location of the ball impact moved away from the node, an increase in amplitude of the higher frequency becomes apparent [Fig. 5(a) and (c)]. In addition, it is quite

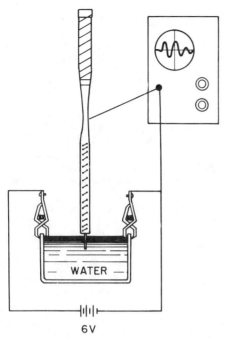

Fig. 4. Method used to determine the position of the racket head as a function of time. The racket handle was clamped (not shown) and the face of the racket hit by a ball. A light-photodetector system (not shown) triggered the scope.

racket and determined the node under those conditions.[5] One would expect the values he obtains to be between the free-free and clamped-free values, but detailed comparisons on similar rackets have not been performed.

Since the hand is a sensitive detector of vibrations, it is possible to get a rough location of the node and the size of the nodal area without any instrumentation. Hold the racket in your hand, face up, and drop a ball onto the strings at various locations. With a ball drop of 20 cm or so you should be able to excite a large enough oscillation to be easily detected.

A series of measurements was made on a number of rackets to determine the location of the node, (with the handle clamped) and the frequencies of oscillation. The data for a few selected rackets are given in Table I. The actual position of the node for a given racket is a function of the mass distribution of the frame and the relative flexibility of various parts of the frame. (Some rackets have stiff shafts and a flexible head, some are the opposite, and some are uniform.) In addition, the flex of the head is influenced by the string tension and pattern, so this can move the node position. To obtain the size of the nodal region, the amplitudes of the oscillation would have to be determined. To do

clear that the phase of the higher-frequency oscillation relative to the fundamental frequency of oscillation changes when going from one side of the node to the other.

The apparatus that was used consisted of a glass beaker filled to the top with tap water, a pinch of salt, two clip leads, a battery, and a fine, stiff wire that was taped to the tip of the racket and the other end connected to an oscilloscope. Less than a centimeter of wire protruded into the water. The scope was triggered by a light source—photodetector system that was interrupted by the ball just before it hit the strings. The ball was suspended on a thread (as a pendulum) and released each time from the same height to give reproducible results.

The free-free and free-clamped experiments should give roughly the same answer for frequency and the node location except that when the racket is clamped, its length is now shorter than the length of the free racket by about 10 cm. This means that these two methods will give slightly different results for the location of the node, but the discrepancy should be no more than 2 or 3 cm. However, since in actual use a racket is neither free nor clamped but hand held, to get a true answer Enoch Durbin of Princeton has mounted an accelerometer on the handle of a hand-held

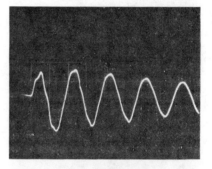

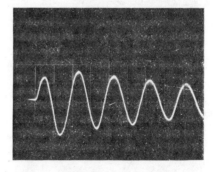

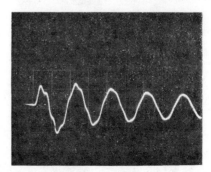

Fig. 5. Typical oscilloscope traces showing racket head position as a function of time. Sweep speed is 20 msec/div.

Table I. Racket oscillation parameters.

Racket	Frequency of fundamental mode (Hz)	Frequency of high mode (Hz)	Distance of node from racket tip (cm)
Prince Pro	36.2	168	13.5
Prince Classic	31.0	127	13.5
Wilson T-2000	26.0	132	13
Head Master	32.5	168	13
Durbin	30.6	138	13
Durbin (with 10 g on tip)	29.2	133	11.5

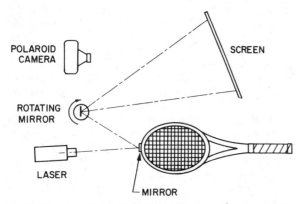

Fig. 6. Alternate method to determine position of racket head as a function of time.

this the resultant output from the liquid potentiometer should be frequency analyzed to get the ratio of the amplitude of the higher mode to the amplitude of the fundamental mode.

The shift in the position of the node was also determined for a single racket when a small mass was taped to the tip. 10 g moved the node approximately 2 cm.

The three definitions of sweet spot all have merit and, in general, the points corresponding to them are located in different places. It is assumed that if the ideal tennis racket could be designed, it would have all three of these points located at the center of the stringed area and have a power region and nodal region covering most of the face of the racket.

ACKNOWLEDGMENTS

I would like to thank Enoch Durbin for several helpful discussions and Jim Hobbs and Oleh Hnatiuk for their help in taking some of the data presented in this paper.

APPENDIX

A second method for observing both the fundamental and higher modes of oscillation with the handle clamped was used and it has the virtue of not requiring an oscilloscope, scope camera, or explanation of why the water beaker system works. (The liquid potentiometer is a pedagogical gem in this author's opinion.) This second method requires a light source (laser), mirror chip glued to the racket tip, rotating mirror, screen, and a camera (Fig. 6). It does require a certain amount of dexterity and coordination to have the ball hit the racket during the time interval when the beam of light is traversing the screen, but with some practice it was possible to get a good picture most of the time.

[1] H. Head, U.S. Patent 3999756, 28 December 1976.
[2] H. Brody, Am. J. Phys. **47**(6), 482 (1979).
[3] F. R. Lacoste, U.S. Patent 3941380, 2 March 1976.
[4] K. Hedrick, R. Ramnath, and B. Mikic, World Tennis **27**(4), 78 (1979).
[5] E. Durbin, patent pending and private communication.

MEDICINE AND SCIENCE IN SPORTS AND EXERCISE
Vol. 15, No. 3, pp. 256-266, 1983

A mechanical analysis of a special class of rebound phenomena

JAMES G. ANDREWS

Division of Materials Engineering and Department of Physical Education (FH),
University of Iowa,
Iowa City, IA 52242

ABSTRACT

ANDREWS, JAMES G. A mechanical analysis of a special class of rebound phenomena. *Med. Sci. Sports Exerc.*, Vol. 15, No. 3, pp. 256-266, 1983. An analytical method is presented for determining the post-impact motion of a rough elastic ball which collides with and rebounds from an arbitrarily oriented rough inertial surface. Pre-impact ball motion is completely general, and the contact interface is assumed to be sufficiently rough to provide a no-slip condition during the restitution phase of the short impact interval. A solution to the impulse-momentum equations is obtained by using Newton's linear coefficient of restitution, and by introducing a torsional coefficient of restitution to account for changes in the magnitude and direction of the component of the ball's angular velocity vector perpendicular to the inertial surface. This method is used to analyze the hop or hook-serve used in advanced-level handball play. An expression for the hop angle γ is derived, which depends on the components of the mass center velocity and angular velocity vectors imparted to the ball by the server. These results are consistent with the natural tendency for right- and left-handed servers to generate characteristic hops to the left and right, respectively. These same results also indicate, however, that many handball authorities are not giving proper instructions when teaching hop servers how the ball should be spinning after hand impact.

IMPACT, IMPULSIVE TORQUE, LINEAR AND TORSIONAL
COEFFICIENTS OF RESTITUTION, HANDBALL, HOP SERVE

Many games and sport activities involve the use of a rough elastic ball as part of the required equipment. During the activity, this ball often strikes and rebounds from a rough inertial surface (e.g., a basketball with the floor and/or backboard; a tennis ball with the court surface; a handball or racquetball with the walls, ceiling, and floor). The post-impact motion of the ball will, of course, depend on its pre-impact motion, the degree of surface roughness at the contact interface, and the impulsive constitutive behavior of the ball when striking the inertial surface.

The classical theory of impact, which predicts the post-impact motions of rough elastic bodies, was presented in its currently accepted form by Routh in 1905 (10). This theory idealizes the collision as occurring during a very short time interval, with both bodies moving as rigid bodies before and after contact. In addition, a common tan-

gent plane (CTP) is assumed to exist in the contact region, and contact is regarded as occurring at a single point. An impulsive force is transmitted to each body at this contact point, and the linear coefficient of restitution, together with the linear and angular impulse-momentum equations of Newtonian rigid body mechanics, are used to determine the post-impact motions of the colliding bodies.

Routh's (10) classical procedure has two serious shortcomings when applied to rough elastic ball rebound situations of the type considered here. The first shortcoming is that this procedure can be quite complex, particularly if all possible frictional effects are considered. To overcome this problem, Routh introduced a clever graphical technique incorporating the motion of a representative point. Despite the introduction of this artifice, the classical procedure remains burdensome and therefore unappealing, particularly if other less complicated procedures can produce results that are in agreement with observation and experiment.

The second shortcoming in the classical analysis (10) is that it assumes point contact occurs. In the ball rebound situations considered here, significant ball deformations often occur during impact, and the contact region therefore becomes appreciable. This finite contact region can transmit not only an impulsive force to the ball, but also an impulsive torque component perpendicular to the CTP. This impulsive torque component can significantly alter the ball's pre-impact angular velocity component perpendicular to the CTP, and lead to a heretofore unconsidered source of ball energy loss during impact. In addition, if the ball subsequently strikes other arbitrarily oriented and rough inertial surfaces, any changes in the magnitude and/or direction of the ball's angular velocity vector due to impact may grossly modify the ball's overall trajectory.

The purposes of this paper are i) to present a simplified theoretical method for determining the post-impact motion of a rough elastic ball that deforms significantly when it collides with and rebounds from a rough inertial surface,

Submitted for publication February, 1982.
Accepted for publication January, 1983.

and ii) to illustrate this method by using it to analyze the hop or hook serve used in advanced-level handball play.

PRELIMINARY CONSIDERATIONS

The ball, in its undeformed state, is assumed to be either a solid sphere of outer radius r, or a spherical shell of outer radius r and constant thickness t. The ball material is assumed to be homogeneous, isotropic, and elastic. Hence, in the undeformed state, the ball's mass-center G and its geometric center coincide, and its physical properties are independent of both location and direction. The ball's elasticity implies that any deformation caused by external forces acting on the ball during impact will disappear when those external forces are no longer present.

During the short time interval when the ball is in contact with the rough inertial surface S, it first deforms and then recovers its initial shape. Let I_c denote the time interval of contact $[t_1, t_3]$, where t_1 is the time of initial contact and t_3 is the time when contact ceases. The short impact interval I_c is commonly subdivided into an initial deformation interval followed by a restitution interval. Let I_d denote the deformation interval $[t_1, t_2]$, where t_2 is the time when deformation stops and restoration of shape begins. Finally, let I_r denote the restitution interval $[t_2, t_3]$.

The ball is assumed to be undeformed when contact begins, and to completely regain its undeformed spherical shape when contact ceases. Any internal vibrations produced by the collision are assumed to be negligible, and the ball is regarded as a rigid body both before and after impact. Its pre- and post-impact motion is, therefore, completely determined by specifying the value of $\bar{v}_G(t)$, the velocity of its mass-center point G relative to the inertial surface S at time t, and by specifying the value of $\bar{\omega}(t)$, the angular velocity of the ball relative to S at time t.

The general ball rebound or impact problem may now be posed in the following way. Determine the values of the ball's mass-center velocity and angular velocity at time t_3 if the values of these two vector quantities are known at time t_1.

SOLUTION METHODOLOGY

The solution to this general impact problem is obtained by making two key assumptions about the nature of the impact, by writing the standard linear and angular impulse-momentum equations of Newtonian rigid-body mechanics, and by solving this system of equations using two experimentally obtainable ball-rebound parameters—Newton's standard linear coefficient of restitution, and a somewhat analogous torsional coefficient of restitution.

The first assumption is that the impact produces a non-negligible contact area at time t_2, and this rough contact area transmits not only a three-dimensional impulsive force, but also an impulsive torque directed perpendicular to the CTP. The six scalar linear and angular impulse-

momentum equations written for the ball during I_c, therefore, contain four scalar kinetic unknowns—the three components of the impulsive contact force, and the single component of the impulsive contact torque.

The second assumption is that the contact interface between ball and inertial surface is sufficiently rough to destroy any tendency for the ball to slide on the surface during I_d, and to cause a no-slip condition to persist during I_r. This assumption leads to two scalar constraint equations that must be satisfied by the six scalar kinematic unknowns, which describe the ball's motion at time t_3.

The two experimentally obtainable ball-rebound parameters, the linear and torsional coefficients of restitution, are defined by two independent scalar equations, which impose additional requirements on the kinematic unknowns describing the ball's motion at time t_3. Their simultaneous use implies that another assumption must be satisfied; namely, that the maximum linear and torsional ball deformations both occur at time t_2.

The general impact problem, therefore, contains 10 scalar unknowns—the three components of $\bar{v}_G(t_3)$, the three components of $\bar{\omega}(t_3)$, the three components of the impulsive contact force during I_c, and the single component of the impulsive contact torque during I_c. This problem is solved by writing the six scalar linear and angular impulse-momentum equations for the ball during I_c, by writing the two scalar, no-slip constraint equations for the ball at time t_3, and by appending the two scalar equations relating the ball-rebound parameters to the ball's kinematic unknowns at time t_3.

Before considering the general case, it is instructive to examine a simple special case that illustrates the class of rebound situations covered by this method, and why this method is needed.

Special Case. Consider the case when a rough imperfectly elastic ball (e.g., a racquetball) is dropped from a rest height h above a rough horizontal inertial surface S while spinning about a vertical axis through its center G. Let R:Qxyz be a right-handed orthogonal coordinate system fixed in S with origin at point Q where contact begins, and with the z-axis perpendicular to S and directed vertically upward through G. Let $\bar{i}$, $\bar{j}$, and $\bar{k}$ denote unit vectors associated with the positive x-, y-, and z-axes of R, respectively. When impact begins, the ball's mass-center velocity and angular velocity are equal to $v_{1z}\bar{k}$ and $\omega_{1z}\bar{k}$, respectively, where v_{1z} is negative and ω_{1z} is arbitrarily taken to be positive.

After impact, the ball is observed to rebound to a height d above S, where d is always less than h. In addition, for certain balls and for sufficiently rough contact surfaces, the ball is observed to be rotating about the z-axis in a direction opposite to its direction before impact, and at a reduced spin rate! Thus, when contact ceases, the ball's mass-center velocity and angular velocity are $v_{3z}\bar{k}$ and $\omega_{3z}\bar{k}$, respectively, where ω_{3z} is negative and smaller in magnitude than $|\omega_{1z}|$, and v_{3z} is positive and can be de-

258

termined in terms of the approach speed v_{1z} and the linear coefficient of restitution e_l from the equation

$$e_l = -v_{3z}/v_{1z}. \qquad [1]$$

In this case, the ball's behavior during impact is similar to that of a combined dissipative linear and dissipative torsional spring. The linear spring analogy accounts for the change in the algebraic sign of the z-component of $\bar{v}_G$, with the loss of translational kinetic energy a function of e_l^2 and evidenced by d being less than h. In the same sense, the ball may also be regarded as a dissipative torsional spring during impact. At time t_1 the torsional spring is undeformed, and during I_d it compresses about the z-axis until at time t_2 the ball's rotational motion about the z-axis stops. During I_r, the torsional spring recovers and finally resumes its initial undeformed configuration when contact ceases. The torsional-spring analogy thus accounts for the change in the algebraic sign of the ball's angular velocity component in the z-direction.

The mechanical energy dissipated by the torsional spring mechanism during impact is evidenced by the fact that $|\omega_{3z}|$ is less than $|\omega_{1z}|$. The rotational kinetic energy lost due to this reduction in spin rate is equal to $(\frac{1}{2})mk^2[(\omega_{1z})^2 - (\omega_{3z})^2]$, where k is the ball's diametral radius of gyration. If the ball material is linearly elastic in both compression and torsion, and if no slip occurs during contact, then the ball rebounds such that not only does d equal h, e_l equal 1, and v_{3z} equal $-v_{1z}$, but ω_{3z} is equal to $-\omega_{1z}$. This observation suggests a simple and straightforward way to account for the mechanical energy lost due to the ball's dissipative torsional deformation about the z-axis during impact.

Let e_t denote a *torsional coefficient of restitution* equal to the ratio of the z-components of the resultant angular impulses acting on the ball about G during I_r and I_d. The no-slip assumption during I_r, and in particular the requirement that both $\bar{v}_G(t_2)$ and $\bar{\omega}(t_2)$ vanish, leads to an expression for e_t of the form

$$e_t = -\omega_{3z}/\omega_{1z}. \qquad [2]$$

Thus, if the ball is elastic and non-hysteretic in torsion as well as in direct compression, and if no slip occurs during contact, then e_t is equal to 1 and equation 2 gives the desired result that ω_{3z} is equal to $-\omega_{1z}$. Noting that ω_{3z} and ω_{1z} are of opposite algebraic sign, it is clear that e_t is a non-negative number which is never greater than unity (i.e., $0 \leq e_t \leq 1$).

The translational kinetic energy lost during this special impact situation may be expressed as

$$mg(h-d) = (\frac{1}{2})m[(v_{1z})^2 - (v_{3z})^2] = mgh(1 - e_l^2).$$

The rotational kinetic energy lost during impact is similarly found to be equal to

$$(\frac{1}{2})mk^2[(\omega_{1z})^2 - (\omega_{3z})^2] = (\frac{1}{2})mk^2(\omega_{1z})^2(1 - e_t^2).$$

Because the total mechanical energy lost during impact is merely the sum of these two expressions, its value can be obtained if the values of mg, d, h, k, ω_{1z}, and ω_{3z} are known. Consequently, an experimental test situation corresponding to this special case of impact may be used to determine e_t based on equation 2. The value of e_t so de-

termined therefore reflects the amounts of rotational kinetic energy lost due to both torsional hysteresis and contact friction in this special case of direct central impact with normal spin.

General case. Let a rough imperfectly-elastic ball strike a rough inertial surface S such that when contact begins, the ball's mass-center velocity $\bar{v}_G(t_1)$ has a component perpendicular to S and a component parallel to S. Referring to Figure 1, let a right-handed orthogonal coordinate system $R:Qxyz$ be fixed in S with the origin at the surface point Q, which contacts the ball first. Let the z-axis be perpendicular to S and be directed away from S through G at time t_1. Let the x-axis be chosen in the same direction as the component of $\bar{v}_G(t_1)$ parallel to S. For this orientation of R,

$$\bar{v}_G(t_1) = v_{1x}\bar{i} + v_{1z}\bar{k}, \qquad [3]$$

where v_{1x} is positive and v_{1z} is negative. Let the ball's angular velocity at time t_1 be denoted by $\bar{\omega}(t_1)$. For the most general case of ball motion prior to contact,

$$\bar{\omega}(t_1) = \omega_{1x}\bar{i} + \omega_{1y}\bar{j} + \omega_{1z}\bar{k}, \qquad [4]$$

where the initial spin components ω_{1x}, ω_{1y}, and ω_{1z} are of arbitrary magnitude and algebraic sign.

Let

$$\bar{v}_G(t_3) = v_{3x}\bar{i} + v_{3y}\bar{j} + v_{3z}\bar{k} \qquad [5]$$

and

$$\bar{\omega}(t_3) = \omega_{3x}\bar{i} + \omega_{3y}\bar{j} + \omega_{3z}\bar{k} \qquad [6]$$

denote the ball's mass-center velocity and angular velocity vectors when contact ceases. If P is the point in the ball that contacts the surface S at point Q initially, then equations 3 and 4 indicate that

$$\bar{v}_P(t_1) = (v_{1x} - r\omega_{1y})\bar{i} + (r\omega_{1x})\bar{j} + v_{1z}\bar{k}.$$

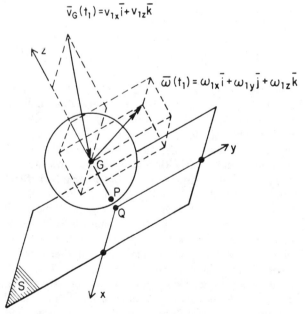

Figure 1—General case of a rough elastic ball colliding with a rough inertial surface S.

Assuming that any initial tendency for P to slide along S is destroyed during I_d, and that a no-slip condition prevails during I_r, then when contact ceases, both the x- and the y-components of $\bar{v}_P(t_3)$ must vanish. Because equations 5 and 6 lead to

$$v_P(t_3) = (v_{3x} - r\omega_{3y})\bar{i} + (v_{3y} + r\omega_{3x})\bar{j} + v_{3z}\bar{k},$$

this assumption requires that v_{3x} must be equal to $r\omega_{3y}$, and that v_{3y} must equal $-r\omega_{3x}$. These two requirements, together with the impulse-momentum equations for the ball during I_c, lead to the following results.

$$v_{3x} = v_{1x} - \left(\frac{k^2}{r^2 + k^2}\right)(v_{1x} - r\omega_{1y}) \quad [7]$$

$$v_{3y} = -\left(\frac{k^2}{r^2 + k^2}\right)r\omega_{1x} \quad [8]$$

$$\omega_{3x} = \left(\frac{k^2}{r^2 + k^2}\right)\omega_{1x} \quad [9]$$

$$\omega_{3y} = \omega_{1y} + \left(\frac{r}{r^2 + k^2}\right)(v_{1x} - r\omega_{1y}) \quad [10]$$

The definitions of the experimentally obtainable linear and torsional coefficients of restitution determine v_{3z} and ω_{3z}. Thus,

$$v_{3z} = -e_l v_{1z}, \quad [11]$$

and

$$\omega_{3z} = -e_t \omega_{1z}. \quad [12]$$

When equations 7–12 are substituted into equations 5 and 6, the two sought-after vectors $\bar{v}_G(t_3)$ and $\bar{\omega}(t_3)$, which describe the motion of the ball when contact ceases, are obtained. These vectors clearly depend on the ball's motion at time t_1 ($\omega_{1x}, \omega_{1y}, \omega_{1z}, v_{1x}, v_{1z}$), the experimentally obtainable linear and torsional coefficients of restitution (e_l, e_t), and the two ball parameters r and k.

The mechanical energy lost during impact can be determined using the initial conditions and equations 7–12. The resulting calculation shows that this lost mechanical energy can be expressed as

$$(\tfrac{1}{2})m\{v_{1z}^2(1 - e_l^2) + k^2\omega_{1z}^2(1 - e_t^2) +$$
$$\left(\frac{k^2}{r^2 + k^2}\right)[(v_{1x} - r\omega_{1y})^2 + (r\omega_{1x})^2]\}. \quad [13]$$

Recalling the expression for $\bar{v}_P(t_1)$, the total mechanical energy lost during impact is seen to depend not only on the initial tendency of P to penetrate into S (as measured by v_{1z}) and slide along S (as measured by $v_{1x} - r\omega_{1y}$ and by $r\omega_{1x}$), but also on the initial tendency of the ball to turn about the z-axis with respect to S (as measured by ω_{1z}). If the ball is elastic and non-hysteretic in both direct compression and torsion, and if there is no initial tendency for P to slide along S, then equation 13 implies there will be no energy lost during impact. Even if the ball is elastic and non-hysteretic in both compression and torsion, equation 13 shows there will be some mechanical energy lost during impact due to friction if there is any initial tendency for P to slide along S in the x- and/or y-directions. Thus, if v_{1x} is not equal to $r\omega_{1y}$ and/or if $r\omega_{1x}$ is not equal to zero, then there will be some mechanical energy lost due to friction as P slides along S during I_d.

APPLICATION TO THE HANDBALL HOP SERVE

One of the most effective serves used in advanced-level handball play is the so-called hop or hook serve. To execute this serve, the server hits the ball quite hard when it is close to the floor, and the ball's trajectory remains close to the floor while the ball is in play. Referring to Figure 2 where the right-to-left serve is viewed from above, the ball leaves the server's hand at A and hits low on the front wall at B. It then rebounds from the front wall and strikes the floor behind the short line at C. When the ball rebounds off the floor at C, it "hooks" or "hops" away from its previous path BC, either "diving" toward the juncture between the side wall and floor (D), or "hopping" back toward the unwary receiver (E). To be most effective, this serve should exhibit a significant hop or deflection away from its path after it strikes the floor at C, and it should never reach the side wall or back wall (in four-wall handball) before it is out of play.

Players who have mastered the hop serve have usually done so only after much experimentation and practice. The really effective hop serves are, therefore, almost the exclusive property of the very best tournament players. These experts and other handball authorities give conflicting advice when explaining how to serve a hop. The ma-

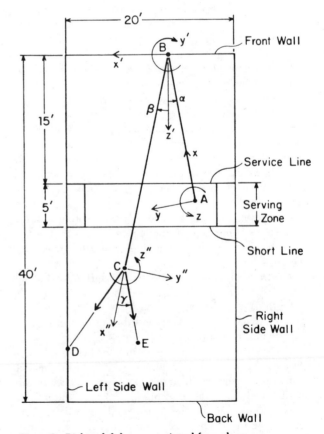

Figure 2—Right-to-left hop serve viewed from above.

jority indicate that the hop is achieved when the server imparts spin to the ball about a *vertical* axis (1,2,4-6,8, 9,15). Other experts, however, indicate that hops are achieved by imparting spin to the ball about an almost *horizontal* axis (3,7,12,14).

In order to use the theory presented previously, the following assumptions will be made: i) the handball and the handball-court surfaces will be regarded as clean and dry and sufficiently rough to destroy any initial tendency for the ball to slide along the court surface (at B and C) during I_d, and to cause a no-slip condition to persist throughout I_r; ii) the ball's initial tendencies to penetrate the court surface and twist about an axis normal to the surface are destroyed during I_d and are partially restored during I_r in accordance with the values predicted using the linear and torsional coefficients of restitution.

To facilitate the analysis of ball motion between contacts with inertial surfaces, the aerodynamic resistance force will be treated as negligible, and the only force acting on the ball between contacts is, therefore, the weight force $m\overline{g}$. The principles of linear and angular momentum then imply that only the vertical component of the ball's mass-center velocity vector $\overline{v}_G(t)$ changes, and the ball's angular velocity vector $\overline{\omega}(t)$ remains constant between contacts. Such a motion is known as simple projectile motion, and whatever the values of $\overline{v}_G$ and $\overline{\omega}$ imparted by either the server at A, or arising due to impacts at B and C, these same values, with the single exception of the vertical component of $\overline{v}_G$, will persist until the ball arrives at its next point of impact.

The mechanical analysis of the hop serve is further facilitated by using three right-handed orthogonal coordinate systems. Restricting the presentation initially to serves from right-to-left, as shown in Figure 2, the first coordinate system, R_1:Axyz, is fixed in the floor with origin at point A directly below the point where the served ball leaves the server's hand. The z-axis is directed vertically upward, and the x-axis is in the same direction as the initial horizontal component of $\overline{v}_G$. The second coordinate system, R_2:Bx'y'z', is fixed in the front wall with origin at the point B where the ball first touches that surface. The z'-axis is perpendicular to the front wall and is directed into the playing area, while the x'-axis is in the same direction as the component of $\overline{v}_G$, which is parallel to the front wall when front wall contact begins. The third coordinate system, R_3:Cx''y''z'', is fixed in the floor with origin at the point C behind the short line where the served ball strikes the floor and the hop is observed. The z''-axis is directed vertically upward, and the x''-axis is in the same direction as the horizontal component of $\overline{v}_G$ just before the ball hits the floor at C.

Let t_o denote the time when the served ball leaves the server's hand, t_1 the time when the ball contacts the front wall at B initially, t_2 the time when front wall contact ceases, t_3 the time when the ball contacts the floor at C initially, and t_4 the time when floor contact ceases. The vertical (z,x)-plane at A contains both $\overline{v}_G(t_o)$ and $\overline{v}_G(t_1)$,

and this plane makes an angle α with the vertical plane containing the horizontal z'-axis at B. The vertical (z'',x'')-plane at C contains both $\overline{v}_G(t_2)$ and $\overline{v}_G(t_3)$, and it makes an angle β with the vertical plane containing the horizontal z'-axis at B. These two angles, α and β, are shown in Figure 2, and they will be referred to as the angles of front-wall incidence and reflection, respectively.

INITIAL CONDITIONS

The server will be assumed to strike the ball in such a way that the direction of $\overline{\omega}(t_o)$ is arbitrary, while $\overline{v}_G(t_o)$ complies with the requirement that the ball leaves the server's hand with considerable speed. This means that, with respect to the unit-vectors $\overline{i}$, $\overline{j}$, and $\overline{k}$ associated with x-, y-, and z-axes of R_1, respectively,

$$\overline{v}_G(t_o) = v_{ox}\overline{i} + v_{oz}\overline{k}$$

and

$$\overline{\omega}(t_o) = \omega_{ox}\overline{i} + \omega_{oy}\overline{j} + \omega_{oz}\overline{k}.$$

For hop serves, the initial horizontal velocity component v_{ox} will be a large positive number, and the initial vertical velocity component v_{oz} will be a small positive number. The initial spin components ω_{ox}, ω_{oy}, and ω_{oz} of $\overline{\omega}(t_o)$ are of arbitrary magnitude and algebraic sign.

At time t_1 when the ball arrives at point B, only the vertical component of $\overline{v}_G$ has changed. Thus,

$$\overline{v}_G(t_1) = \overline{v}_G(t_o) - \mu\overline{k} = v_{ox}\overline{i} + (v_{oz} - \mu)\overline{k},$$

and

$$\overline{\omega}(t_1) = \overline{\omega}(t_o),$$

where μ is a small positive number due to the gravity effect acting on the ball during the short time interval $[t_o,t_1]$. The z-component of $\overline{v}_G(t_1)$ is, therefore, slightly less than v_{oz}, the small positive z-component of $\overline{v}_G(t_o)$. For the sake of simplicity, the vertical component of $\overline{v}_G(t_1)$ will be assumed to be equal to zero. This implies that $\overline{v}_G(t_1)$ is directed entirely in the x-direction, and is therefore horizontal. As a result,

$$\overline{v}_G(t_1) = v_{ox}\overline{i}$$

and

$$\overline{\omega}(t_1) = \omega_{ox}\overline{i} + \omega_{oy}\overline{j} + \omega_{oz}\overline{k}.$$

This not only agrees with what is normally observed for effective hop serves, but it has the additional advantage of allowing for a particularly simple orientation of the R_2:Bx'y'z' coordinate system in the front wall. Because the x'-axis is in the same direction as the component of $\overline{v}_G(t_1)$ parallel to the front wall, the x'-axis will now be horizontal and directed to the left for the case of serves from right to left as shown in Figure 2.

FRONT-WALL IMPACT

The angle of incidence α is selected by the server, and is a function of the server's position in the serving zone and the position of point B, the imaginary target on the front wall. To avoid shadow serves, i.e., serves where the receiver's view of the ball is obstructed by the server's body, α must be greater than some small angle.

Knowing the values of $\bar{v}_G(t_1)$ and $\bar{\omega}(t_1)$, the general case results presented previously may now be used to determine the values of $\bar{v}_G(t_2)$ and $\bar{\omega}(t_2)$ when front-wall contact ceases. To do this it is first convenient to express the pre-impact vectors $\bar{v}_G(t_1)$ and $\bar{\omega}(t_1)$ in terms of their components in the $R_2:Bx'y'z'$ coordinate system. Let $\bar{i}'$, $\bar{j}'$, and $\bar{k}'$ be the unit vectors associated with the positive x'-, y'-, and z'-directions of R_2, respectively, and let

$$\bar{v}_G(t_1) = v_{1x'}\bar{i}' + v_{1y'}\bar{j}' + v_{1z'}\bar{k}'$$

and

$$\bar{\omega}(t_1) = \omega_{1x'}\bar{i}' + \omega_{1y'}\bar{j}' + \omega_{1z'}\bar{k} .$$

The pre-impact values at B can then be written as

$$\begin{aligned}
v_{1x'} &= v_{ox}\sin\alpha; & \omega_{1x'} &= \omega_{oy}\cos\alpha + \omega_{ox}\sin\alpha; \\
v_{1y'} &= 0; & \omega_{1y'} &= -\omega_{oz}; & [14] \\
v_{1z'} &= -v_{ox}\cos\alpha; & \omega_{1z'} &= \omega_{oy}\sin\alpha - \omega_{ox}\cos\alpha.
\end{aligned}$$

Reference to Figure 2 also shows that the angle of incidence α satisfies the relation

$$\tan\alpha = (v_{1x'})/(-v_{1z'}) .$$

In a similar manner, let the post-impact kinematic quantities be denoted by

$$\bar{v}_G(t_2) = v_{2x'}\bar{i}' + v_{2y'}\bar{j}' + v_{2z'}\bar{k}'$$

and

$$\bar{\omega}(t_2) = \omega_{2x'}\bar{i}' + \omega_{2y'}\bar{j}' + \omega_{2z'}\bar{k}' . \qquad [15]$$

Using the general case results to determine the six unknown scalar coefficients in equation 14 gives

$$\begin{aligned}
v_{2x'} &= v_{1x'} - \lambda v_{P1x'}; & \omega_{2x'} &= \lambda\omega_{1x'}; \\
v_{2y'} &= -\lambda r\omega_{1x'}; & \omega_{2y'} &= \omega_{1y'} + rk^{-2}\lambda v_{P1x'}; & [16] \\
v_{2z'} &= -e_l v_{1z'}; & \omega_{2z'} &= -e_t\omega_{1z'};
\end{aligned}$$

where

$$\lambda = k^2/(r^2 + k^2) ,$$

and   [17]

$$v_{P1x'} = v_{1x'} - r\omega_{1y'} .$$

The reflection angle β may be expressed in terms of the components of $\bar{v}_G(t_2)$. Referring to Figure 2,

$$\tan\beta = \frac{v_{2x'}}{v_{2z'}} = \frac{v_{1x'} - \lambda v_{P1x'}}{-e_l v_{1z'}},$$

where $v_{1z'} < 0$, and $v_{P1x'}$ represents the velocity component in the x'-direction at time t_1 of the point P in the ball which first contacts the front wall at point B. If $v_{P1x'}$ vanishes, then the front wall exerts no friction force component on the ball in the x'-direction and $\tan\beta$ is equal to $(\tan\alpha)/e_l$. If the ball is also linearly elastic in compression, then e_l is equal to 1 and α is equal to β.

FLOOR IMPACT

Consider next the contact at C which begins at time t_3. The values of $\bar{v}_G(t_3)$ and $\bar{\omega}(t_3)$ are the same as $\bar{v}_G(t_2)$ and $\bar{\omega}(t_2)$, respectively, except for a change in the vertical component of $\bar{v}_G$. Thus, let

$$\bar{v}_G(t_3) = \bar{v}_G(t_2) + \delta\bar{j}' = v_{2x'}\bar{i}' + (v_{2y'} + \delta)\bar{j}' + v_{2z'}\bar{k}' ,$$

and

$$\bar{\omega}(t_3) = \bar{\omega}(t_2) ,$$

where δ is a small positive number that arises due to the gravity effect acting on the ball during the time-interval $[t_2, t_3]$. In order to determine the values of $\bar{v}_G(t_4)$ and $\bar{\omega}(t_4)$ when contact at C terminates, it is convenient to first express $\bar{v}_G(t_3)$ and $\bar{\omega}(t_3)$ in terms of their components in the $R_3:Cx''y''z''$ coordinate system. Let $\bar{i}''$, $\bar{j}''$, and $\bar{k}''$ be unit vectors associated with the positive x''-, y''- and z''-axes of R_3, respectively, and let

$$\bar{v}_G(t_3) = \bar{v}_{3x'}\bar{i}'' + v_{3y'}\bar{j}'' + v_{3z'}\bar{k}''$$

and

$$\bar{\omega}(t_3) = \omega_{3x'}\bar{i}'' + \omega_{3y'}\bar{j}'' + \omega_{3z'}\bar{k}'' .$$

The pre-impact values at C can then be expressed as

$$\begin{aligned}
v_{3x''} &= v_{2z'}\cos\beta + v_{2x'}\sin\beta; \\
\omega_{3x''} &= \omega_{2z'}\cos\beta + \omega_{2x'}\sin\beta; \\
v_{3y''} &= v_{2z'}\sin\beta - v_{2x'}\cos\beta; & [18] \\
\omega_{3y''} &= \omega_{2z'}\sin\beta - \omega_{2x'}\cos\beta; \\
v_{3z''} &= -v_{2y'} - \delta; & \omega_{3z''} = -\omega_{2y'} .
\end{aligned}$$

In a similar manner, let the post-impact quantities at C be expressed as

$$\bar{v}_G(t_4) = v_{4x''}\bar{i}'' + v_{4y''}\bar{j}'' + v_{4z''}\bar{k}''$$

and   [19]

$$\bar{\omega}(t_4) = \omega_{4x''}\bar{i}'' + \omega_{4y''}\bar{j}'' + \omega_{4z''}\bar{k}'' .$$

Using the general-case results leads to the following expressions for the six unknown coefficients in equation 19:

$$\begin{aligned}
v_{4x''} &= v_{3x''} - \lambda v_{P3x''}; & \omega_{4x''} &= \lambda\omega_{3x''}; \\
v_{4y''} &= -\lambda r\omega_{3x''}; & \omega_{4y''} &= \omega_{3y''} + rk^{-2}\lambda v_{P3x''}; & [20] \\
v_{4z''} &= -e_l v_{3z''}; & \omega_{4z''} &= -e_t\omega_{3z''};
\end{aligned}$$

where λ is defined in equation 17, and

$$v_{P3x''} = v_{3x''} - r\omega_{3y''} \qquad [21]$$

represents the velocity component in the x''-direction at time t_3 of the point P in the ball which first contacts the floor at point C.

HOP ANGLE γ

The characteristic and most significant property of the hop serve is the ball's sideways jump away from the vertical (z'',x'')-plane after it leaves the floor at C. Let the hop angle γ be the angle between the vertical (z'',x'')-plane and the vertical plane containing $\bar{v}_G(t_4)$. Reference to Figure 2 then shows that

$$\tan\gamma = v_{4y''}/v_{4x''} . \qquad [22]$$

The problem now becomes one of expressing $\tan\gamma$ in terms of the components of $\bar{v}_G(t_o)$ and $\bar{\omega}(t_o)$ and then determining how, for serves, from right to left, the hop angle γ will vary with changes in the components of $\bar{\omega}(t_o)$.

Substituting equations 21, 20, 18, 17, 16, and 14, into 22 leads to the following expression:

$$\tan\gamma = \frac{-\omega_{oH}[\cos(\alpha-\beta+\phi_o) + \zeta\cos\beta\cos(\alpha+\phi_o)]}{L - \omega_{oH}[\sin(\alpha-\beta+\phi_o) - \zeta\sin\beta\cos(\alpha+\phi_o)]} . \qquad [23]$$

The terms ω_{oH}, ϕ_o, ζ and L appearing in equation 23 are defined as follows. The term ω_{oH} denotes the magnitude of the horizontal component of the ball's angular velocity vector at time t_o. Thus, if $\bar{\omega}_{oH}$ represents the horizontal component of $\bar{\omega}(t_o)$, then reference to Figure 3 indicates that

$$\bar{\omega}(t_o) = \bar{\omega}_{oH} + \omega_{oz}\bar{k} ,$$

262

where

$$\overline{\omega}_{oH} = \omega_{ox}\overline{i} + \omega_{oy}\overline{j} = \omega_{oH}(\cos\phi_o\overline{i} + \sin\phi_o\overline{j}),$$

and

$$\omega_{oH} = |\overline{\omega}_{oH}| = (\omega_{ox}^2 + \omega_{oy}^2)^{1/2}.$$

The symbol ϕ_o denotes the angle between the vertical (z,x)-plane at A and the vertical plane containing $\overline{\omega}(t_o)$ as shown in Figure 3. Thus,

$$\tan\phi_o = \omega_{oy}/\omega_{ox}.$$

The term ζ is a ball parameter defined by the expression

$$\zeta = \frac{e_t - \lambda}{\lambda} = \frac{r^2 e_t - k^2(1 - e_t)}{k^2}, \qquad [24]$$

where r and k are the ball's outer radius and diametral radius of gyration, respectively, and e_t is the ball's torsional coefficient of restitution.

The symbol L in equation 23 is defined by the expression

$$L = \left(\frac{r}{k^2}\right) v_{ox}\left[\left(\frac{e_l}{\lambda}\right)\cos\alpha\cos\beta + \left(\frac{r}{k}\right)^2\sin\alpha\sin\beta\right]$$
$$+ \left(\frac{r}{k}\right)^2\omega_{oz}\sin\beta, \qquad [25]$$

where e_l is the ball's linear coefficient of restitution. Consequently, L depends on v_{ox} and ω_{oz}, and is independent of ω_{oH}. For the standard handball under standard conditions, e_l, r, and t are approximately equal to 0.80, 2.35 cm, and 0.65 cm, respectively, where t is the thickness of the ball's spherical shell. Hence,

$$k^2 = \frac{2[r^5 - (r - t)^5]}{5[r^3 - (r - t)^3]} = 2.85 \text{ cm}^2.$$

Recalling equation 17, the value of λ is approximately equal to 0.34. Substituting these approximate values into equation 25 yields

$$L = 0.82 v_{ox}[(2.35)\cos\alpha\cos\beta + (1.94)\sin\alpha\sin\beta]$$
$$+ (1.94)\omega_{oz}\sin\beta,$$

where the units of L will be in $\text{rad}\cdot\text{s}^{-1}$ if v_{ox} is in $\text{cm}\cdot\text{s}^{-1}$ and if ω_{oz} is in $\text{rad}\cdot\text{s}^{-1}$.

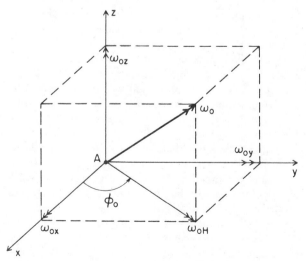

Figure 3—Definition of $\phi_o = \tan^{-1}(\omega_{oy}/\omega_{ox})$.

In order to determine the relative influence of ω_{oH} and ω_{oz} on the values of $\tan\gamma$ as given by equation 23, note first from Figure 2 that the sum of α and β must vary from a small positive value (just large enough to avoid shadow serves where the server's body screens the receiver from seeing the ball until it passes the server's body), to a maximum value of approximately 60°. Assuming, from experience, that α and β are approximately equal for hop serves, it is therefore clear that α (or β) may vary from a small positive angle to a maximum positive angle of about 30°. Letting α equal β in equation 23 leads to the result that

$$\tan\gamma = \frac{-\omega_{oH}[\cos\phi_o + \zeta\cos\alpha\cos(\alpha + \phi_o)]}{L' - \omega_{oH}[\sin\phi_o + \zeta\sin\alpha\cos(\alpha + \phi_o)]},$$

where

$$L' = 0.82 v_{ox}[(1.94) + (0.41)\cos^2\alpha] + (1.94)\omega_{oz}\sin\alpha \qquad [26]$$

denotes the value of L when β equals α.

Let the server impart a given amount of spin ω_o to the ball at time t_o, where

$$|\overline{\omega}(t_o)| = \omega_o = (\omega_{ox}^2 + \omega_{oy}^2 + \omega_{oz}^2)^{1/2}$$
$$= (\omega_{oH}^2 + \omega_{oz}^2)^{1/2}. \qquad [27]$$

Also assume that the server can, without undue difficulty, orient the ball's initial spin axis in any desired direction. Thus, ω_o may be imparted only about the vertical z-axis, or solely about a horizontal axis, or about some axis which is neither vertical nor horizontal. It is clear from equation 27 that if the server makes ω_{oz} equal to ω_o, then ω_{oH} vanishes. In this case, equation 23 indicates that the hop angle γ will be equal to zero regardless of the initial spin rate ω_o. Consequently, if the server initially spins the ball solely about the vertical z-axis, the theory presented in this paper predicts that γ vanishes, or there will be no hop at C!

If the server makes ω_{oH} equal to ω_o, then ω_{oz} vanishes and equation 23 indicates that, depending upon the orientation (ϕ_o) of the initial horizontal angular velocity vector $\overline{\omega}_{oH}$, the hop angle γ can be positive, negative, or zero. Hence, if the server spins the ball solely about a horizontal axis, the theory predicts that there are angular orientations ϕ_o of $\overline{\omega}_{oH}$ such that γ does *not* vanish and the ball therefore hops at C!

APPROXIMATE ANALYSIS

If the server imparts all of the initial spin ω_o about an axis which is neither vertical nor horizontal, then a more complex situation arises. In order to determine if this choice of initial spin axis orientation will theoretically produce larger hop angles γ than those obtained when ω_{oH} is equal to ω_o, it is necessary to first obtain approximate values for ζ, ω_o, and L as defined by equations 24, 26, and 27, respectively. For this purpose and using conservative estimates, let e_t be taken to lie between 0.34 and 0.68. Using the previously calculated value of 0.34 for λ, equation 24 then indicates that ζ lies between 0 and 1. Further, consider as "typical" the case when both α and β are equal to 15°, v_{ox} is equal to 2000 cm·s⁻¹, and ω_o is equal to 125 rad·s⁻¹. Replacing ω_{oz} in equation 26 by $\pm(\omega_o{}^2 - \omega_{oH}{}^2)^{\frac{1}{2}}$ as obtained from equation 27, the "typical" value of L′ then becomes

$$L'_{typ} = 3800 \pm 0.50\,(\omega_o{}^2 - \omega_{oH}{}^2)^{\frac{1}{2}},$$

where the term $\pm\,0.50\,(\omega_o{}^2 - \omega_{oH}{}^2)^{\frac{1}{2}}$ in L'_{typ} varies from zero when ω_{oH} is equal to ω_o, to 62.5 rad·s⁻¹ when ω_{oH} vanishes. Hence, irrespective of the initial spin direction, L'_{typ} is a large positive number.

EXTREME VALUES OF THE HOP ANGLE γ

The typical hop angle γ_{typ} for right-to-left serves may therefore be expressed in the form

$$\tan\gamma_{typ} =$$
$$\frac{-\omega_{oH}[\cos\phi_o + \zeta(0.97)\cos(\alpha+\phi_o)]}{3800 \pm 0.50(\omega_o{}^2 - \omega_{oH}{}^2)^{\frac{1}{2}} - \omega_{oH}[\sin\phi_o - \zeta(0.26)\cos(\alpha+\phi_o)]},\quad[28]$$

where ζ lies between 0 and 1. An examination of this expression reveals that, due to the large positive value of v_{ox}, the denominator is always a large positive quantity which is essentially insensitive to the choice of ω_{oH} in the given range. The numerator term, however, depends directly on the value of ω_{oH}, and non-trivial hop angles γ_{typ} will occur only when ω_{oH} does not vanish. If equation 28 is therefore regarded as expressing γ_{typ} as a function of the single variable ω_{oH} on the closed interval [0, 125] rad·s⁻¹, and if use is made of the theory of relative maxima and minima for functions of one variable from the calculus, then it can be shown that γ_{typ} has no *relative* maxima or minima in the given ω_{oH} interval.

The maximum value of γ_{typ} must therefore occur when ω_{oH} is either equal to zero or equal to ω_o. Because γ_{typ} vanishes when ω_{oH} is equal to 0, the greatest hop angles must occur when ω_{oH} is equal to ω_o, or when ω_{oz} vanishes. The server will therefore obtain the greatest hop at C when all the available spin is imparted solely about a horizontal axis!

In order to determine how γ varies with the angular orientation ϕ_o of $\overline{\omega}_{oH}$, equation 23 will be considered for the simplified case when α equals β and L is regarded as a large positive constant that is much greater than ω_{oH} (i.e., $L \gg \omega_{oH}$). Equation 23 then takes the form

$$\tan\gamma = \frac{-\omega_{oH}[\cos\phi_o + \zeta\cos\alpha\cos(\alpha+\phi_o)]}{L - \omega_{oH}[\sin\phi_o - \zeta\sin\alpha\cos(\alpha+\phi_o)]},\quad[29]$$

where ζ lies between 0 and 1, and α is a positive angle less than 15°. A plot of the variation of γ with ϕ_o (for serves from right to left), as expressed by equation 29, is shown in Figure 4. Radial lines drawn from the center of the circle at A to points on the circumference represent possible orientations of $\overline{\omega}_{oH}$ as a function of ϕ_o. Vectors emanating from C represent corresponding orientations of the horizontal component of $\overline{v}_C(t_4)$. Numbered points on the circle's circumference correspond to like-numbered vectors emanating from C.

Figure 4 indicates that the ball will hop to the right at C in the positive y″-direction if the horizontal component

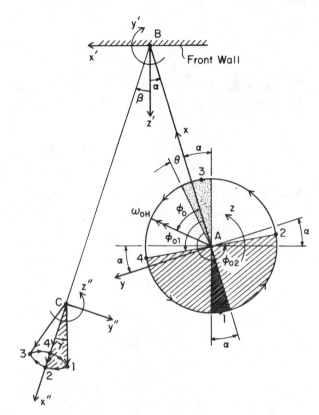

Figure 4—Variation in the hop angle γ with ϕ_o for a right-to-left hop serve.

$\overline{\omega}_{oH}$ of the initial angular velocity vector $\overline{\omega}(t_o)$ lies in the rearward cross-hatched portion of the circle of radius ω_{oH} centered at A. The angle ϕ_{oR} which produces the greatest hop to the right at C depends on the values of L, ω_{oH}, α, and ζ. For the conditions considered here (i.e., $L \gg \omega_{oH}$ and $0 \le \zeta \le 1$), it can be shown that ϕ_{oR} lies between $(180° - \alpha)$ and 180°, and corresponds to point 1 in Figure 3. Thus, if $\overline{\omega}_{oH}$ is directed to the rear at A and lies somewhere within the double cross-hatched portion of the circle, the server will theoretically achieve the greatest hop to the right at C.

If the ball is struck behind and below G, then this clockwise spin, as viewed from behind the server, about an axis almost parallel to the horizontal x-axis should be easier for a right-handed server to apply because the right-hander's natural swing is inward and across the front of the body.

Figure 4 also shows that the ball will hop to the left at C in the negative y″-direction if $\overline{\omega}_{oH}$ lies in the forward portion of the circle centered at A. The angle ϕ_{oL}, which produces the greatest hop to the left at C, also depends on the values of L, ω_{oH}, α and ζ. For the conditions considered here, it can be shown that ϕ_{oL} corresponds to point 3 in Figure 4 and lies between $(-\alpha)$ and θ, where $\sin\theta$ is equal to (ω_{oH}/L). Because $L \gg \omega_{oH}$, θ is a small angle. Thus, if $\overline{\omega}_{oH}$ is directed forward at A and lies somewhere

264

within the dotted portion of the circle, the server will theoretically achieve the greatest hop to the left at C.

If the ball is struck behind and below G, then this clockwise spin, as viewed from behind the server, about an axis almost parallel to the horizontal x-axis should be easier for a right-handed server to apply because the right-hander's natural swing is inward and across the front of the body.

The theory predicts that there are two orientation angles ϕ_{o1} and ϕ_{o2} of $\overline{\omega}_{oH}$ that will theoretically produce no hop at C. These angles are found by setting the numerator of equation 29 equal to zero. This gives

$$\tan \phi_{o1} = \frac{1 + \zeta \cos^2\alpha}{\zeta \sin \alpha \cos \alpha}.$$

For values of ζ between 0 and 1, it can be shown that θ_{o1} lies between $(90° - \alpha)$ and $90°$ and corresponds to point 4 in Figure 4, whereas ϕ_{o2} is equal to ϕ_{o1} plus $180°$, and corresponds to point 2. Figure 4 also shows that as ω_{oH} rotates about A in the counterclockwise sense, the projection of the terminus of $\overline{v}_G(t_4)$ on the (x'',y'')-plane moves along its closed path in the clockwise sense.

LEFT-TO-RIGHT SERVES

For serves from left to right, a similar analysis leads to an expression for the hop angle γ in the form

$$\tan \gamma = \frac{-\omega_{oH}[\cos (\alpha - \beta - \phi_o) + \zeta \cos \beta \cos (\alpha - \phi_o)]}{L + \omega_{oH}[\sin (\alpha - \beta - \phi_o) - \zeta \sin \beta \cos (\alpha - \phi_o)]} \quad [30]$$

The similarities between equations 30 and 23 suggest that the same reasoning applied to equation 23 may be used on equation 30 to discover the best orientation for the initial spin axis to produce the greatest hops at C for serves from left to right. This leads to the conclusion that the greatest hops at C will occur when all of the available spin ω_o is again imparted to the ball solely about a horizontal axis!

To obtain the variation in γ due to changes in ϕ_o, consider the simplified case of equation 30 when α is equal to β and L is regarded as a large positive constant much greater than ω_{oH}. Equation 30 then takes the form

$$\tan \gamma = \frac{-\omega_{oH}[\cos \phi_o + \zeta \cos \alpha \cos (\alpha - \phi_o)]}{L - \omega_{oH}[\sin \phi_o + \zeta \sin \alpha \cos (\alpha - \phi_o)]}, \quad [31]$$

where ζ lies between 0 and 1, and α is a positive angle less than $15°$. A plot of the variation in γ with ϕ_o (for serves from left to right) as expressed by equation 31, is shown in Figure 5. This sketch indicates that the ball will hop to the right at C in the positive y''-direction if ω_{oH} lies in the rearward cross-hatched portion of the circle of radius ω_{oH} centered at A. The angle ϕ_{oR}, which produces the most pronounced hop to the right at C, depends on the values of L, ω_{oH}, α, and ζ. For the conditions considered here, it can be shown that ϕ_{oR} (point 3 in Figure 5) lies between $(180° - \theta)$ and $(180° + \alpha)$, where $\sin \theta$ is a small value equal to (ω_{oH}/L). Thus, the server will theoretically obtain the greatest hop to the right at C if $\overline{\omega}_{oH}$

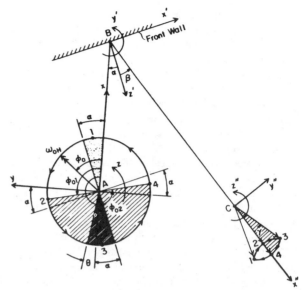

Figure 5—Variation in the hop angle γ with ϕ_o for a left-to-right hop serve.

is directed to the rear and lies in the double cross-hatched portion of the circle. This counterclockwise spin, as viewed from behind the server, about an axis almost parallel to the horizontal x-axis should be easier for the left-handed server to apply for the same reasons as noted previously.

Figure 5 also shows that the ball will hop to the left at C in the negative y''-direction if $\overline{\omega}_{oH}$ lies in the forward portion of the circle of radius ω_{oH} centered at A. The angle ϕ_{oL}, which produces the greatest hop to the left at C, also depends on the values of L, ω_{oH}, α, and ζ. For the conditions considered here, it can be shown that ϕ_{oL} (point 1 in Figure 5) lies between 0 and α. Thus, the server theoretically achieves the greatest hop to the left at C if $\overline{\omega}_{oH}$ is directed forward and lies in the dotted portion of the circle at A. When viewed from behind the server, the clockwise spin about an axis almost parallel to the horizontal x-axis should be easier for the right-handed server to apply.

The theory again predicts that there are two orientation angles, ϕ_{o1} and ϕ_{o2} of $\overline{\omega}_{oH}$, which theoretically cause no hop at C. These two angles are obtained by setting the numerator of equation 31 equal to zero. This gives

$$\tan \phi_{o1} = -\frac{1 + \zeta\cos^2\alpha}{\zeta \sin\alpha \cos\alpha}.$$

For values of ζ between 0 and 1, it can be shown that ϕ_{o1} (point 2 in Figure 5) lies between $90°$ and $(90° + \alpha)$, whereas ϕ_{o2} (point 4 in Figure 5) is equal to ϕ_{o1} plus $180°$.

DISCUSSION

Figures 4 and 5 indicate that, for serves from either right to left or left to right, the theoretical results are consistent. The most pronounced hops are achieved when the initial spin axis is horizontal and almost parallel to the x-axis.

When viewed from behind the server, the theory predicts in both cases that the greatest hop to the right at C is obtained with counterclockwise spin. Similarly, in both cases, the greatest hop to the left at C is theoretically achieved with clockwise spin about a line that is almost parallel to the horizontal x-axis when viewed from behind the server. If the initial spin axis is vertical, or if it is horizontal and almost parallel to the y-axis, then the theory predicts in both cases that there will be no hop off the floor at C.

The preceding theoretical results conflict with the instructional advice offered by many of handball's leading authorities (1,2,4-6,8,9,15). They suggest that the server should attempt to spin the ball about a vertical axis, whereas the theory predicts that *no* hop will be observed if the server imparts spin solely about a vertical axis.

Two writers (11,13) have suggested that the initial spin axis should be oriented at some angle that is neither vertical nor horizontal. The theory, however, predicts that for a given initial spin magnitude, the hop will be reduced if the initial spin axis is not oriented entirely in the horizontal plane.

Those experts (3,7,12,14) who indicate that the initial spin should be imparted about a horizontal axis, as when you turn a door knob, provide advice that is consistent with the theoretical analysis. Even here, however, the only investigator who attempts to analyze the situation (14) states that this horizontal spin is not affected by front-wall impact. This statement is in direct conflict with the requirement that a *reversal* in horizontal spin direction must occur during front-wall impact in order for the natural and reverse hops to occur as they do at first floor contact.

The theory also indicates that hops to the left should be easier for a right-handed server to achieve, while hops to the right should be more easily produced by left-handed servers. This agrees with the common court observation that the right-hander's "natural" hop is to the left, whereas the left-hander's "natural" hop is to the right.

This qualitative consistency between theory and observation is reinforced by further arguments that support the predicted results and underscore the importance of the underlying theory. In this regard, consider again the ball-floor impact at C. An appeal to Newton's laws leads to the conclusion that, in order to produce any hop at C, the ball must have a pre-impact angular velocity component parallel to the x''-axis. Thus, if $\omega_{3x''}$ is positive, then $v_{4y''}$ should be negative, and if $\omega_{3x''}$ is negative, then $v_{4y''}$ should be positive.

If the prior ball-front wall impact at B is then examined, particularly for the case when both α and β are negligibly small, then it is clear that in order to have a post-impact

angular velocity component in the x'' or z'-direction, the ball must have a pre-impact angular velocity component in this same direction. However, if the observed "natural" hops of right- and left-handers are to be predicted by the theory, then the front-wall impact at B must produce a change in the algebraic sign of the ball's angular velocity component in the z'-direction! This reversal in the direction of the normal spin component implies that the ball must behave like a torsional spring during front-wall impact. Such a reversal in the direction of normal spin cannot be accounted for properly by the classical analysis procedure of Routh (10) wherein point contact is assumed. It can, however, be successfully accounted for by making the assumptions incorporated in the theoretical analysis presented in this paper, and by introducing the concept of a torsional coefficient of restitution.

Finally, it should be emphasized that the theory presented in this paper may be generalized and used to solve a broader class of impact problems provided (i) an impulsive torque acts perpendicular to the CTP, (ii) the terminal no-slip assumption is satisfied, and (iii) the linear and torsional coefficients of restitution are known and are appropriate for the particular impact problem under consideration. For example, if a spinning tennis ball is struck by a moving tennis racquet, and if the impulsive force and torque applied to the racquet handle by the player during ball impact are known, then the preceding theory can be used to predict the immediate post-impact motions of both the ball and the racquet. The 12 scalar impulse-momentum equations (six for each object) written over the contact interval I_c contain 12 scalar kinematic unknowns (six for each object), together with four scalar kinetic unknowns (the three impulsive force and one impulsive torque components). The four additional scalar equations needed to solve for all 16 scalar unknowns are the two scalar no-slip kinematic constraint conditions when contact ceases and the two scalar equations defining the linear and torsional coefficients of restitution in terms of kinematic quantities.

The theory presented in this paper should not be used to analyze all those situations when a rough elastic ball collides with a rough inertial surface. The two assumptions that most severely restrict the theory's applicability are the no-slip requirement when contact ceases and the stipulation that both the linear and torsional deformations of the ball reach their maximum values simultaneously at time t_2 when I_d ends and I_r begins. If these two requirements are not satisfied, the concept of a torsional coefficient of restitution will not be appropriate, even if the deformations are large and an impulsive torque is transmitted at the contact region.

REFERENCES

1. BONTEMPO, U. Hopping the ball. In: *Championship Handball by the Experts*, J.W. Reznik (Ed.). Cornwall, NY: Leisure Press, 1976, p. 77.

2. FOSTER, M.A. Retired colonel discloses 'secret' of hopping ball without arm soreness. In: *Championship Handball by the Experts*, J.W. Reznik (Ed.). Cornwall, NY: Leisure Press, 1976, pp. 78–81.

266

3. HABER, P. and M. LEVE. *Inside Handball*. Chicago: Reilly and Lee, 1970, p. 19.
4. MAND, C.L. *Handball Fundamentals*. Columbus, OH: Merrill, 1968, pp. 36–38.
5. NELSON, R.C. and H.S. BERGER. *Handball*. Englewood Cliffs, NJ: Prentice-Hall, 1971, p. 24.
6. O'CONNELL, C. *Handball Illustrated*. New York: Ronald Press, 1964, pp. 26–28.
7. PACKER, H. Varying serve to find vulnerable spot. In: *Championship Handball by the Experts*, J.W. Reznik (Ed.). Cornwall, NY: Leisure Press, 1976, pp. 100–102.
8. PHILLIPS, B.E. *Handball: Its Play and Management*. New York: Ronald Press, 1957, pp. 16–18.

9. ROBERSON, R. and H. OLSON. *Beginning Handball*. Belmont, CA: Wadsworth, 1962, pp. 15–18.
10. ROUTH, E.J. *The Elementary Part of a Treatise on the Dynamics of a System of Rigid Bodies*, 7th Edition. New York: Dover, 1960, pp. 260–268.
11. SHAW, J.H. *Handball*. Boston: Allyn and Bacon, 1971, pp. 48–51.
12. TYSON, P. and M. LEVE. *Handball*. Pacific Palisades, CA: Goodyear, 1971, pp. 53–58.
13. YESSIS, M. *Handball*, 1st Edition. Dubuque, IA: Brown, 1966, pp. 29–31.
14. YESSIS, M. *Handball*, 3rd Edition. Dubuque, IA: Brown, 1977, pp. 58–61.
15. YUKIC, T. *Handball*. Philadelphia: Saunders, 1972, pp. 84–91.

J. Biomechanics Vol. 12, pp. 893–904.
© Pergamon Press Ltd. 1979: Printed in Great Britain.

0021-9290/79/1201-0893 $02.00/0

THE INFLUENCE OF TRACK COMPLIANCE ON RUNNING

Thomas A. McMahon and Peter R. Greene

Division of Applied Sciences, Pierce Hall, Harvard University, Cambridge, MA 02138, U.S.A.

Abstract – A model of running is proposed in which the leg is represented as a rack-and-pinion element in series with a damped spring. The rack-and-pinion element emphasizes the role of descending commands, while the damped spring represents the dynamic properties of muscles and the position and the rate sensitivity of reflexes. This model is used to predict separately the effect of track compliance on step length and ground contact time. The predictions are compared with experiments in which athletes ran over tracks of controlled spring stiffness. A sharp spike in foot force up to 5 times body weight was found on hard surfaces, but this spike disappeared as the athletes ran on soft experimental tracks. Both ground contact time and step length increased on very compliant surfaces, leading to moderately reduced running speeds, but a range of track stiffness was discovered which actually enhances speed.

INTRODUCTION

Running is essentially a series of collisions with the ground. As the animal strikes the surface, its muscles contract and ultimately reverse the downward velocity of the body. Intuition argues that a surface of suitably large compliance is bound to change running performance. Running on a diving springboard slows a man down considerably, while running on a trampoline is all but impossible. Our goal in this paper will be to find an analytic expression for the change in the runner's speed, step length and foot contact time as a function of the track stiffness, and to compare these predictions with experiment.

The simplification that the muscles of locomotion and their reflexes act essentially as springs is lent support by recent developments in the study of neural motor control. Reflexes, however, require some time to act – anyone who has unexpectedly stepped off a curb will recall the sharp jolt which results when the antigravity muscles of the leg are not prepared for the impact. Melville Jones and Watt (1971) have shown that approximately 102 msec are required for reflex activity from the otolith apparatus to activate the antigravity muscles in man, so that unexpected falls of less than about 5.0 cm are unaccompanied by reflex accommodation. Even the simple stretch reflex requires a substantial portion of the running step cycle. The latency of EMG changes associated with automatic responses to a change in limb load are found to be in the range of 79 msec for elbow flexion in man (Crago *et al.*, 1976) and near 25 msec for soleus muscles in decerebrate cats (Nichols and Houk, 1976). Since the supported period in human running is typically 100 msec, neither reflexes of vestibular nor stretch origin can be expected to participate in the first quarter of the stance phase, and therefore the antigravity muscles of the leg must be principally under the control of command signals from higher motor centers during this time. In the later portion of the stance phase, however, the stretch reflex can be expected to make important modifications of the efferent activity of α-motorneurons. Houk (1976) has argued that muscle stiffness, rather than muscle length, is the property which is regulated by the stretch reflex. He points out that a competition between length-related excitation contributed by muscle spindle receptors and force-related inhibition contributed by Golgi tendon

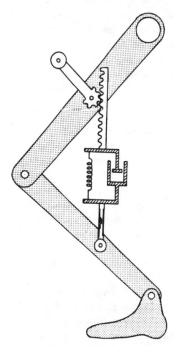

Fig. 1. Schematic representing the separate role of descending commands (rack-and-pinion) and muscle properties plus local reflexes (damped spring). The motion of the rack and pinion element determines the influence of track stiffness on step length. The runner's mass and the damped spring determine the influence of track stiffness on ground contact time.

* *Received 17 July 1978.*

894
THOMAS A. MCMAHON and PETER R. GREENE

organs could result in the ratio of muscle force to length being regulated, rather than either one exclusively. Support for this view comes from ramp stretches of the soleus muscle in decerebrate cats (Nichols and Houk, 1976). These studies show how reflex action can compensate for stretch-induced reductions in muscle force, thus preserving a linear force–length relation in a stretched muscle which would otherwise show acute nonlinearity. Houk suggests that the action of a muscle (or a pair of muscles) about a joint might reasonably be represented as a rack and pinion in series with a spring.

A modification of this scheme is shown in Fig. 1. Movement commands would crank the rack-and-pinion to a new set point for the joint angle, but force disturbances from the outside would deflect the limb by an amount dictated by the damped spring. The dashpot in parallel with the spring is not specifically mentioned in Houk's model, but is necessary to include the rate sensitivity of the stretch receptors and other feedback elements when both muscle force and length are changing rapidly.

Representation of the leg and its musculature as a linear damped spring has already proved successful in describing an exercise in which the subject jumps onto a force platform, falling on the balls of the feet without flexing the knees, and with the ankles forcefully extended (Cavagna, 1970). From the resultant damped oscillation in vertical force (frequency about 3.5 Hz), Cavagna (1970) calculated the effective spring stiffness and damping constant of the extensors of the ankle. The oscillations were always underdamped, with a damping ratio of about 0.2.

In subsequent sections, the function of the damped spring in Fig. 1 is separated from the function of the rack-and-pinion. First, under the assumption that the rack-and-pinion is locked, we treat the vertical motion of the runner as an underdamped mass–spring system, and calculate the time required to rebound from the track as a function of track compliance. The assumption that the rack-and-pinion is locked emphasizes the local control of muscle stiffness at the segmental level during the middle and late portions of the stance phase of limb motion. Later, we assume that the damped spring is locked, and geometric considerations are applied to the rack-and-pinion element to calculate the effect of the track compliance on the man's step length. This assumption emphasizes the pre-programmed, non-reflex control of limb position during the early extension phase, before and just after the foot touches the ground. Finally we obtain a prediction for the top running speed as a function of track compliance. Observations of subjects running on experimental tracks of various stiffness are presented for comparison. Although the calculations show that the man is severely slowed down when the track stiffness is less than his own spring stiffness, there exists an intermediate range of track stiffnesses where his speed is either unaffected by the track or somewhat enhanced.

METHODS

Experiments

Experimental board track. A single-lane running surface 26.25 m in length was constructed of 1.9 cm plywood boards. Each board was 40.6 cm long in the running direction by 121.9 cm wide. The boards were screwed to 4.4 × 8.9 cm rails which served as supports, as shown in Fig. 2. The spring stiffness of the track could be altered by moving the supporting rails closer or farther apart. A typical load–deflection calibration, obtained by applying 0.22 kN weights to a 12.7 cm circular aluminum plate representing the foot, is also shown in Fig. 2(b). The time required for the runner to pass between two transverse light beams 8.20 m apart provided a measure of the runner's speed. The force applied to the track by the runner's foot was measured by a Kissler 9261A force plate, which was linear ±0.5% over a force range of 0–2.0 kN, and had a natural frequency when loaded with a 70 kg man above 200 Hz. A small 60.9 cm square panel of 0.95 cm phenolic resin board supported at either end by 2.54 cm square pine rails rested upon the force plate, as shown. The separation of the 2.54 cm rails was adjusted until the load–deflection curve of the phenolic board matched that of the track to within 2.0%. In this way, the runner was presented with a level track surface of uniform compliance, and the vertical foot force could be measured as he struck the phenolic board. Each subject ran down the center of the track, to ensure that he experienced the compliance measured by the load–deflection calibration. A 16 mm ciné camera, operating at approximately 60 frames per sec, provided a photographic record. A clock in the field of view of the camera was used to calibrate the camera speed.

A total of 8 subjects, all males between the ages of 21 and 34 yr, participated in the experiments (Table 1). The subjects were told to run at a uniform speed. They alternated runs on the track with runs on the concrete surface beside the track. Each subject ran at a variety of speeds, including his top speed. All runners wore conventional running shoes with thin, flat soles.

Pillow track. In order to determine the effect of a very soft surface, the board track was replaced by a 10.9 m long sequence of foam-rubber pillows, each measuring 1.22 m wide by 0.91 m high by 2.74 m long. The runner's speed, step length and ground contact time on each stride were determined by film analysis. The load–deflection curve for the pillow track is shown in Fig. 2(c). There was a large hysteresis, resulting in a different stiffness for loading and unloading at a particular force level.

For the purpose of subsequent calculations, the spring stiffness of the pillows was evaluated at two different force levels. This was done by obtaining the local slope of the force–deflection curve at the 0.8 kN level (1.0g), corresponding to foot forces of the order of the runner's body weight, and at the 1.34 kN level (1.67g), corresponding to the mean foot forces to be

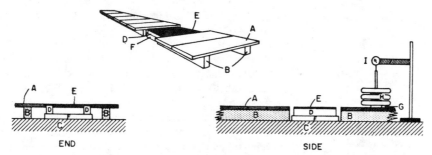

Fig. 2. (a) Three views of the experimental wooden track. A—plywood running surface, B, D–spruce supporting rails, C–concrete floor, E—phenolic resin board, F—force platform, G—aluminum plate representing the foot, H—weights, I—displacement gauge.

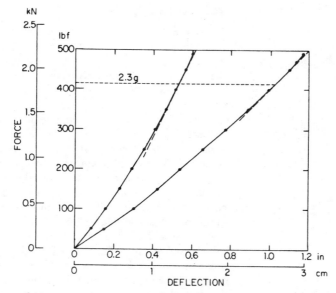

Fig. 2. (b) Force–deflection curves for two configurations of the experimental wooden track. Tangents fit to the 2.3g level give $k_t = 13,333$ lbf/ft (195 kN/m) and 6857 lbf/ft (100 kN/m).

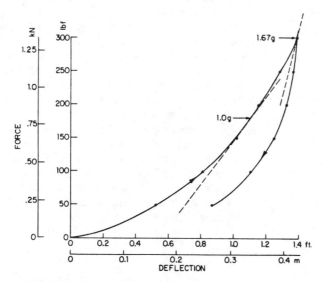

Fig. 2. (c) Force–deflection curves for the foam rubber pillow track, showing 1.0g and 1.67g tangents, which give $k_t = 320$ lbf/ft (4.67 kN/m) and 985 lbf/ft (14.4 kN/m).

896   THOMAS A. MCMAHON and PETER R. GREENE

Table 1. Experimental subjects

Subject	Weight (lbf)	Weight (kN)	Height (m)	Leg length, l (m)	Hard surface Step length, L_o (m)	contact time, t_o (sec)	L_o/l	$\left[\dfrac{(\pi/t_o)^2 l}{(1-\zeta^2)g}\right]$	Runner's spring stiffness $k_m = \dfrac{m_m\pi^2}{t_o^2}/(1-\zeta^2)$ (lbf/ft)	(kN/m)
M.F.	180	0.800	1.93	1.09	0.896	0.108	0.83	135.1	6781	98.9
N.H.	175	0.778	1.91	1.00	0.814	0.100	0.81	144.3	7683	112.1
T.M.	175	0.778	1.91	1.01	0.890	0.136	0.88	78.8	4084	59.6
J.J.	180	0.800	1.83	0.978	0.878	0.112	0.89	112.5	6474	94.5
P.G.	190	0.845	1.93	1.05	0.896	0.120	0.86	105.2	5796	84.6
J.C.	160	0.712	1.79	0.969	0.859	0.109	0.89	117.7	5919	86.4
S.R.	160	0.712	1.78	0.960	0.890	0.122	0.93	93.07	4725	68.9
G.L.	150	0.667	1.78	0.960	0.878	0.131	0.92	80.72	3842	56.1

expected during the foot contact time on this compliant surface. The 1.0g and 1.67g pillow stiffnesses were 4.67 kN/m and 14.38 kN/m, respectively.

Each subject generally was tested at between 5 and 8 different running speeds on each of the four track surfaces (concrete, board track at 195 kN/m, board track at 100 kN/m and pillow track). There were exceptions, as in the case of the pillow track, where only 4 subjects participated. As explained in the next sections, foot contact time t_c was included in the tabulations (for Fig. 8) only at the highest running speed of each runner on each surface (27 points). By contrast; each step length determination required a straight-line fitting process like that shown in Fig. 6. Therefore each of the 27 step length points (Fig. 7) represents 5 or more individual runs.

Theoretical considerations

Foot contact time. As a general principle, cushioning works to decrease the forces between colliding bodies by increasing the time of the collision. Joggers know that they are less prone to ligament injuries and shinsplits when they run on somewhat compliant surfaces such as turf, as opposed to city pavements. Typical stiffnesses of some running surfaces are shown in Table 2.

In Fig. 3, a one-dimensional model of the runner and the track is shown which ignores motion in the forward direction and considers only the vertical component. In this model, we have fixed the rack-and-pinion-element from Fig. 1 in a single position, thus emphasizing the role of muscle reflex stiffness. The mass m_m is the man's mass, and k_m is the lumped spring stiffness of the muscles and reflexes acting to extend his hip, knee and ankle. The effective mass of the track surface (the magnitude of an equivalent mass concentrated at a point) is m_t, and the spring stiffness of the track (the inverse of its compliance) is shown as k_t. In the figure, all the masses and springs are attached, so that only that half-cycle of the motion for positive downward displacements of the man (x_m) and track (x_t) has any correspondence with physical reality. When x_m and x_t are negative, the man's foot would, in the actual situation, be separated from the track surface and would therefore not interact with it. Although the permanent connection of the man to the track is fictitious, it makes the mathematics convenient and corresponds approximately to the real situation during the contact portion of the stepping cycle.

Track mass. Let us ignore the man's damping for the moment, and consider the undamped vibration of the man and the track. The natural frequency ω_n(rad/sec) of the lowest mode of vibration, in which the two masses move downward in phase, is given by Den Hartog (1956):

$$\omega_n^2 = \frac{(m_t + m_m)k_m}{2m_t m_m} + \frac{k_t}{2m_t}$$

$$- \frac{\sqrt{[m_t k_m + m_m(k_t + k_m)]^2 - 4m_t m_m k_t k_m}}{2m_t m_m}. \quad (1)$$

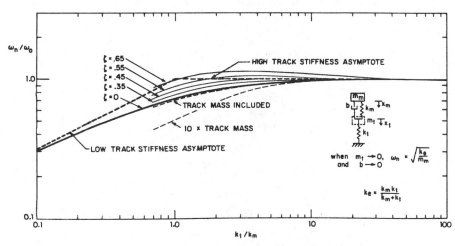

Fig. 3. Normalized natural frequency vs normalized track stiffness. The inset shows the damped two-mass, two-spring system. Heavy line shows zero damping, zero track mass.

Table 2. Stiffness of running surfaces

Material	Stiffness (lbf/ft)	(kN/m)
concrete, asphalt	300,000+	4376
packed cinders	200,000	2918
board tracks	60,000	875
experimental wooden track	13,333	195
experimental wooden track	6857	100
pillow track at 1.67g	985	14.4

In the rigid-track limit, $k_t/k_m \to \infty$, $m_t/m_m \to \infty$, and the above expression becomes $\omega_o^2 = k_m/m_m$. In the remainder of the paper, the subscript o will denote the rigid-track limit. In the limit as the track becomes very soft, $\omega_n^2 = k_t/(m_m+m_t)$. The intersection of these two asymptotic behaviors occurs at a track stiffness $k_t^\dagger = k_m(m_m + m_t)/m_m$, where the natural frequency $\omega_n^\dagger$ is given by

$$\left(\frac{\omega_n^\dagger}{\omega_o}\right)^2 = \frac{m_t + m_m}{m_t} - \frac{\sqrt{m_m^2 + m_t m_m}}{m_t}. \qquad (2)$$

A broken line showing the influence of track mass on frequency is shown in Fig. 3. Assuming a conventional wooden track construction in which a 1.9 cm × 1.22 m × 2.44 m plywood panel reinforced by 4.4 × 8.9 cm stringers is the running surface, the effective mass of the track is 21.0 kg, which makes $m_t/m_m = 0.25$ for an 87 kg runner. In this calculation, the effective mass of the track is obtained by Rayleigh's method, assuming a sinusoidal two-dimensional mode shape (Timoshenko, 1937). Under these circumstances, the result is that the effective mass is half the total mass of the panel.

The solution shown in Fig. 3 including the track mass is seen to be not very different from the solution for the low track mass limit, $\omega_n^2 = k_t k_m/m_m(k_t + k_m)$, plotted as a heavy solid line just above it. For comparison, the solution when the track mass is increased by a factor of 10 is also shown.

Influence of the man's damping. Since the track mass encountered in practice has so little effect on ω_n, we consider it no further. Taking $m_t = 0$, we investigate the combined effect of the force–velocity relation in the man's muscles and the velocity feedback in the man's stretch reflexes, represented here by the dashpot element, b, shown in the schematic drawing in Fig. 3. From the solution presented in Appendix A, the normalized frequency ω_n/ω_o is plotted as a function of dimensionless track stiffness k_t/k_m for 4 choices of the damping ratio $\zeta = b/(2\sqrt{m_m k_m})$. Notice that the damped curves lie above the undamped ones, a consequence of the fact that the dashpot element tends to stiffen the man's impedance in this normalized comparison. The parameter k_m required for this calculation was determined for each ζ from $k_m = m_m \omega_o^2/(1 - \zeta^2)$, where m_m and $t_o = \pi/\omega_o$ are the mass and hard surface contact time appropriate for subject M.F. These curves will later be compared with experimental results.

Step length. Two sequences of stick figures, obtained by analysis of the ciné films, are shown in Fig. 4. Each figure was drawn by connecting points locating the major limb joints. The topmost point locates the position of the ear. A remarkable observation is that the trajectory of the ear, and therefore of the otolith apparatus sensing head acceleration, is relatively level, whether the subject runs on the pillows or on the hard surface. Both lower extremities and a single upper extremity are shown in the stick figures. When the subject runs on the pillows, as shown at the bottom, his stance foot sinks into the foam rubber, but the swing foot always remains above the undeflected pillow surface. The extended leg encounters the pillow surface in a position when hip flexion is greater than is the case for running on a hard surface. The step length on the pillow surface is consequently greater.

This observation may be used to construct a model for the influence of track compliance on step length. In the schematic diagram of Fig. 5(a), the leg, length *l*, is shown with the knee fully extended at the

THOMAS A. McMAHON and PETER R. GREENE

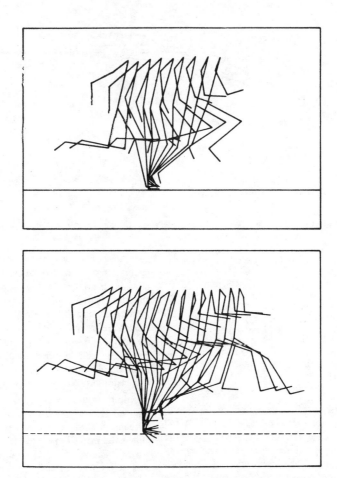

Fig. 4. Stick figures of subject M.F. running, from films. Top: hard surface; bottom: pillow track. Solid line shows undeflected surface of pillow track; broken line shows mean deflection of pillows over an entire step cycle. Only those figures for which the foot was in contact with the surface are drawn. The framing speed was 59 frames/sec.

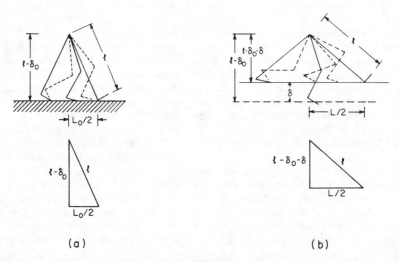

Fig. 5. Schematic of a step on (a) hard surface and (b) pillow track. Solid line shows the stance leg, broken line shows the swing leg moving forward. Because the foot descends a distance δ into the pillows, the step length on the pillow track is necessarily greater.

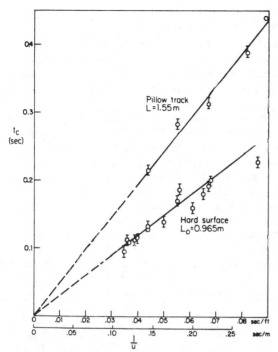

Fig. 6. Ground contact time t_c vs inverse running speed $1/u$, for runner M.F. The straight lines through the origin show that an individual's step length is constant, independent of the running speed, on a particular surface. Step length is greater on the pillow track than on the hard surface. Error bars show maximum uncertainty due to film reading.

moment of contact with the hard surface. It is also shown in mid-stance, when the knee is flexed, and at the end of the stance phase, just before the toe is lifted. In mid-stance, the length of the leg is only $l - \delta_o$, where the shortening δ_o is assumed to be a constant length,

independent of running speed, achieved by the "rack-and-pinion" higher postural controllers for the purpose of maintaining the body (and therefore the ear) on an approximately level trajectory. Notice that this assumption effectively fixes the length of the damped spring in Fig. 1 as if the spring stiffness k_m were now taken to be infinite. Since $\delta_o = 9.6$ cm for subject M.F. running on the hard surface, but the maximum deflection of his "spring" would be expected to be only 1.86 cm, this assumption appears to be justified. The important point is that the base of the triangle shown in the lower part of the figure is longer, and thus the step length L is longer on the pillow surface (Fig. 5b). The distance δ is the mean deflection of the pillow surface throughout a complete stride, including the aerial phases. If the man were not running at all, but merely standing quietly on the pillows, he would be standing in a well of depth $\delta = m_m g / k_t$, where k_t in this instance is the pillow stiffness measured at the 1.0g force level. The broken line in Fig. 5(b), representing the mean deflection of the pillow track, plays the same role as the solid line in Fig. 5(a): all details of the step are arbitrarily presumed to be the same, including the distance from the broken line to the hip, $l - \delta_o$. Only the hip flexion angle at which the heel contacts the track is different on the pillows, leading to the longer step length. Applying the Pythagorean theorem to the triangle in Fig. 5(b),

$$L = 2\sqrt{l^2 - (l - \delta_o - \delta)^2}. \qquad (3)$$

The constant δ_o may be written in terms of the step length on the hard surface, L_o,

$$\delta_o = l - \sqrt{l^2 - L_o^2/4}. \qquad (4)$$

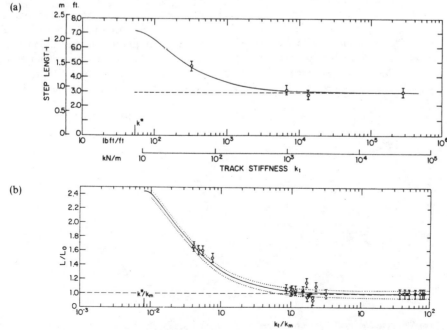

Fig. 7. Step length vs track stiffness. The solid line shows the theoretical prediction. (a) Subject M.F. alone. (b) Dimensionless plot showing all 8 subjects.

900   THOMAS A. MCMAHON and PETER R. GREENE

Combining equations (4) and (3), with $\delta = m_m g/k_t$,

$$L = 2\sqrt{l^2 - [(l^2 - L_o^2/4)^{1/2} - m_m g/k_t]^2}. \quad (5)$$

Equation (5) is plotted in Fig. 7(a), assuming a 0.8 kN man with a leg length $l = 1.09$ m and a hard-surface step length $L_o = 0.896$ m (appropriate for subject M.F.). When the expression in the square bracket is zero, the step length has reached its maximum, namely twice the leg length. Thus running on surfaces whose stiffness is less than $k_t^* = m_m g/\sqrt{l^2 - L_o^2/4}$ would not be possible, according to this model, since the hips would have descended below the surface of the pillows.

RESULTS

Dimensionless plotting

Since the results are presented on dimensionless axes, we have included a short justification for the validity of this procedure in Appendix B. Basically, the method is required because we wish to compare the performances of several runners on the same figure. If the dimensional axes were retained, the performance of a single runner could be compared to a single line especially computed for that runner (for example, Fig. 7a for subject M.F.), but a complete presentation of the results would require as many figures as there were runners.

Man's spring determined by foot contact time

In plotting each data point on a typical dimensionless graph (e.g. Fig. 7b), it was first necessary to know the man's spring stiffness k_m. This, in general, is a function of the man's effort, and increases as he runs faster. In Figs. 3, 7, 8 and 9, we compare only the maximum running performance as a function of track stiffness and therefore $k_m = m_m \omega_o^2/(1-\zeta^2)$ where $\omega_o = \pi/t_o$, t_o is the time the foot is in contact with the ground while running at maximum effort on the hardest surface, and ζ is the damping ratio (assumed to be 0.55 for each runner, as explained below).

Foot contact time vs track stiffness

In Fig. 8, foot contact time t_c/t_o is plotted against track stiffness k_t/k_m. The theoretical line represents a damping ratio for the man of $\zeta = 0.55$. This damping ratio was chosen among the four shown in Fig. 3, on the basis of its satisfactory fit to the experimental points shown in Fig. 8. The linearized spring stiffness of the pillows was taken as the 1.67g stiffness, 14.4 kN/m, since the pillows acted with this stiffness during most of the time the runner was in contact with the track, when foot forces were in the range of 1.67 times body weight. We shall return to this point later, with an explanation of how the figure 1.67g was determined.

The error bars for each point show the estimated maximum uncertainty in reading the films and force records, which was generally less than $\pm 7.0\%$. The dotted lines on either side of the theoretical line are displaced by one standard deviation σ, where σ is estimated from the root of the mean of squared residuals:

$$\sigma \simeq [\hat{\sigma}^2]^{1/2} = \left[\frac{1}{n-2}\sum_{i=1}^{n}[y_i(x)-y(x)]^2\right]^{1/2}. \quad (6)$$

Here $y_i(x)$ is the measured value and $y(x)$ is the computed value of a parameter at a given x (Meyer, 1975).

Step length independent of running speed on a given surface

In their comprehensive study of human gait, Cavagna *et al.* (1976) noticed that the step length, the distance a man travels while one foot is in contact with the ground, is a constant value for a given individual runner on a hard surface, independent of his running speed. We were able to corroborate this finding, as shown in Fig. 6 for subject M.F. The time in contact with the ground, t_c, is proportional to the inverse of velocity. The slope of this line defines the step length L, which is independent of the speed but very much larger on the pillows than on the hard surface. When L/L_o is plotted against k_t/k_m in Fig. 7, a very good agreement is

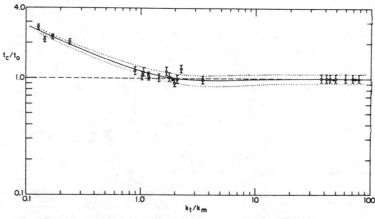

Fig. 8. Normalized foot contact time t_c/t_o vs normalized track stiffness, assuming damping ratio $\zeta = 0.55$. Open circles show data produced by film analysis; closed circles show force platform data. Error bars show limits of uncertainty due to film and oscilloscope reading; dotted lines are one standard deviation above and below the theoretical line.

found with the theoretical line, with a standard deviation of 0.045.

Note that the spring stiffness used for the pillows is now the 1.0g stiffness, 4.67 kN/m, because the deflection δ in equation (3) must correspond to the distance a man would sink down if he were merely standing at rest on the (linearized) pillows. Recall that the step length theory was derived entirely on the basis of geometrical considerations, and did not involve the man's spring stiffness. It is therefore consistent with the observed fact that the step length on a particular surface is independent of running speed.

Foot force

The average vertical force applied to the ground by the foot during a step is equal to the runner's mass times his mean vertical acceleration,

$$\bar{F} = m_m g + 2 m_m v/t_c, \qquad (7)$$

where v is the downward vertical velocity at the moment of contact. In our experiments, we measured v by integrating the force over the duration of t_c, and found no significant difference in v for a given subject on a hard, as opposed to a compliant, surface. Thus v is taken to be a constant, found for a particular runner from the area under the force–time curve,

$$v = \frac{\int_o^{t_c} (F - m_m g)\mathrm{d}t}{2 m_m}. \qquad (8)$$

Taking a representative $v = 0.732$ m/sec for a 0.8 kN subject, and using values for t_c obtained from Fig. 8 in the case where the damping ratio $\zeta = 0.55$, a dimensionless $\bar{F}/\bar{F}_o$ vs k_t/k_m curve may be plotted (Fig. 9). This theoretical line agrees reasonably well with the force-plate data points, and shows that no appreciable change in the mean levels of foot force can be expected until the track stiffness is significantly less than the stiffness of the man. Note that it was not possible to

measure foot force during the pillow running experiments, but the prediction would be that average force was lowered to 0.71 times its hard-surface value, or about 1.67 times body weight.

Representative force signatures, traced from the oscilloscope photographs for subject J.C., are shown in the lower portion of Fig. 9. On the hard surface, the *initial* contact of the foot with the ground produced a spike in foot force which often exceeded 5 times body weight. This spike was either absent or very much attenuated when the same subject ran on a compliant track. We suspect this dramatic reduction of foot force at initial contact is the reason that all subjects reported a subjective impression of increased running comfort on the compliant surfaces relative to the hard surfaces.

Running speed

Having obtained predictions for the ground contact time t_c and step length L, we may put these together to obtain the running speed, $u = L/t_c$. A consequence of the fact that L and t_c are nearly constant in the intermediate range of track stiffness is that running speed should not be significantly affected until k_t/k_m drops below 1.0. At low track stiffness, foot contact time t_c increases, but so does step length, so the runner is not slowed down as much as contact time alone would predict. For example, running on the pillows increased t_c by an average factor of 2.3, but the runner's speed was not halved. Instead, since step length increased by a factor of 1.6, the runner's speed was preserved at 70% of its hard-surface value.

DISCUSSION

Limitations of the analysis

It is important to review the assumptions made at various points throughout this paper, and to understand how they limit the analysis. We began by representing the antigravity muscles and their reflexes

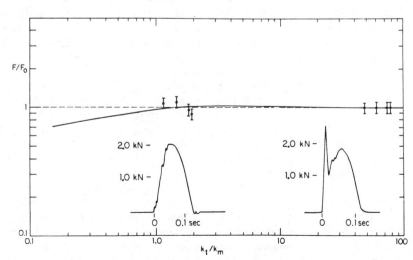

Fig. 9. Normalized average foot force vs normalized track stiffness. Solid line shows theory, solid points show average force platform results for each of four subjects. Insets show how the initial force transient experienced on hard surfaces is abolished on the experimental wooden track.

by the simple mechanical system shown in Fig. 1, and used the dynamic characteristics of the spring and dashpot to calculate the influence of track stiffness on ground contact time. Later, we used the conceptual model of the rack-and-pinion element, ignoring the damped spring, to calculate the influence of track stiffness on step length. We acknowledged the intrinsic nonlinearity of the force–length characteristic of stretched muscle, but claimed, following Houk, that reflex compensation acts to restore linearity. In another paper (Greene and McMahon, 1979), we have measured the short-range spring stiffness of the muscular reflexes of the leg, as a function of both knee angle and total force, and find that the effective spring stiffness of the leg varies by a factor of 2 over the knee angles encountered in running, but most of the variation occurs in the first 15 degrees of knee flexion. Remarkably, the spring stiffness at a constant knee angle is found to be no more than 25% greater as the subject carries loads up to twice body weight on his shoulders. Thus, as long as the knee angle θ is kept within the range $15° \leq \theta \leq 45°$, as it commonly is during running, our assumption of one single spring stiffness for the leg throughout the step cycle is reasonably valid.

Another simplification involves the pillows: we have assumed that their load–deflection curve is linear, whereas Fig. 2 shows that it is most distinctly nonlinear. We have also neglected damping in the pillows, which is probably not entirely justified. In addition, we have neglected the horizontal compliance of the pillows.

We dealt with the nonlinearities of the pillows by assigning the 1.0g stiffness its proper role in determining step length, while we assumed that the 1.67g stiffness determines foot contact time. Since we found that most of the subjects applied a sustained vertical force of about 2.4 times body weight to the hard-surface track, according to Fig. 9, $\bar{F}$ on the pillows should be about 0.7 times this value, or about 1.67 times body weight. Thus our measurements and predictions are consistent with our basic assumptions about foot force on the pillows.

Man's stiffness

The concept of the man's spring stiffness is complicated by the fact that it depends on his effort. Under the assumptions of equation (1), when the man runs twice as fast, his spring stiffness increases by a factor of 4. We have attempted to eliminate the effort dependence of the man's stiffness in this paper by making comparisons of performance on different track stiffnesses only when the man is running at maximal effort.

Man's damping

We employed a damping element in parallel with the man's spring because we knew that (1) isolated muscles obey a Hill force–velocity curve (Hill, 1938), and (2) the muscle spindles return velocity information to the spinal cord. Our decision about how much damping

was realistic depended on the curve-fitting procedure shown in Fig. 3. Since the curve representing a damping ratio of $\zeta = 0.55$ provided the best fit through the experimentally determined points for t_c/t_o, we took that value of ζ for subsequent calculations of foot force and running speed.

Our assumption that the damping element is linear is certainly a great oversimplification. Katz (1939) showed 40 years ago that the damping parameter b (Fig. 3) is about 6 times greater for slow lengthening as opposed to slow shortening in isolated muscles. The extent to which this effect is modified by reflex phenomena is unknown.

Could an independent set of experiments, not involving running, be proposed to measure the value of ζ appropriate for running? Cavagna (1970) was able to measure ζ by allowing his subjects to go through several damped cycles of ringing while the muscles of the calf remained in sustained contraction. In running, no such ringing oscillations could ever be observed because the foot remains in contact with the ground for only half a ringing cycle, and the total mechanical energy is the same at the beginning and the end of each supported period. In fact, this is a property of all nonlinear oscillations, that the energy lost in the dissipative mechanism matches the energy added per cycle by the "negative resistance" phenomenon. Thus, only indirect techniques which change the operating characteristics of the oscillator by changing one of its component parts (here we used the track) can serve to analyze the remaining components.

As a final remark it is worth noting that the model of the vertical motion of the runner shown in Fig. 3 can easily be made into a nonlinear oscillator. Suppose that when both the man and the track are descending, and therefore when the leg is being flexed by the man's downward momentum, the damping constant of the dashpot, b, is positive, as was assumed in the body of the paper. As an additional feature, suppose that when the trajectory of the center of mass x_m reaches its lowest point, b suddenly switches sign, and provides negative damping for the next half-cycle. The sudden change in the sign of b requires a sudden advance in the phase of $x_m - x_t$ with respect to x_m, and this requires a step change in the length of x_t. The essential result is that, by postulating a damping which switches sign at mid-stride (as if it were determined by joint receptors), we may generate an oscillatory motion whose amplitude does not decay with time, and yet whose period is the same as the simple system with linear damping discussed in the body of the paper.

SUMMARY AND CONCLUSIONS

Beginning with a model of the antigravity muscles and reflexes which assumes that they have an automatic, or reactive, component which makes them behave like a damped linear spring, and this is in series with a purposeful component which behaves like an externally controlled rack-and-pinion, we have derived ground contact time, step length, foot force and running speed as functions of track compliance. These predictions are compared with the results of experiments in which subjects ran alternately on a compliant and a hard surface, and the agreement is generally good.

Very compliant tracks, which have a spring stiffness much less than the man's stiffness, are responsible for a marked penalty in the runner's performance. For example, when a man runs on a track which is 0.15 times his own stiffness, his running speed is reduced to 0.70 times the speed he could run on a hard surface.

On tracks of intermediate compliance, the analytical model predicts a slight speed enhancement, due to a decrease in foot contact time and an increase in step length, by comparison with running on a hard surface. Another important advantage of such tracks of intermediate compliance is the marked attenuation of the early peak in foot force, which can reach 5.0 times body weight in running on a hard surface.

A permanent indoor track having a stiffness about three times the man's stiffness has recently been completed in the new indoor athletic facility at Harvard University. Experience to date indicates that good runners are able to better their usual times in the mile by about 5 sec on this track. This represents a speed enhancement of 2%, in good agreement with the theoretical prediction. The runners also report that this track is particularly comfortable to run on, and is apparently responsible for a very low rate of running injuries.

Acknowledgements – This work was supported in part by the Harvard University Planning Office and by the Division of Applied Sciences, Harvard University, Cambridge, Massachusetts.

REFERENCES

Bridgeman, P. W. (1931) *Dimensional Analysis*. Yale University Press, New Haven.

Cavagna, G. A. (1970) Elastic bounce of the body. *J. appl. Physiol.* **29**, 279–282.

Cavagna, G. A., Thys, H. and Zamboni, A. (1976) The sources of external work in level walking and running. *J. Physiol., Lond.* **262**, 639–657.

Crago, P. E., Houk, J. C. and Hasan, Z. (1976) Regulatory actions of human stretch reflex. *J. Neurophysiol.* **39**, 925–935.

Greene, P. R. and McMahon, T. A. (1979) Reflex stiffness of the antigravity muscles. *J. Biomechanics* **12**

Den Hartog, J. P. (1956) *Mechanical Vibrations*, p. 90. McGraw-Hill, New York.

Hill, A. V. (1938) The heat of shortening and the dynamic constants of muscle. *Proc. R. Soc.* (B) **126**, 136–195.

Houk, J. C. (1976) An assessment of stretch reflex function. *Prog. Brain Res.* **44**, 303–313.

Katz, B. (1939) The relation between force and speed in muscular contraction. *J. Physiol.* **96**, 45–64.

Melville Jones, G. and Watt, D. G. D. (1971) Observations on the control of stepping and hopping movement in man. *J. Physiol.* **219**, 709–727.

Meyer, S. L. (1975) *Data Analysis for Scientists and Engineers*, p. 393. Wiley, New York.

Nichols, T. R. and Houk, J. C. (1976) Improvement in linearity and regulation of stiffness that results from actions of stretch reflex. *J. Neurophysiol.* **39**, 119–142.

Timoshenko, S. (1937) *Vibration Problems in Engineering*. Van Nostrand, New York.

NOMENCLATURE

b linear dashpot damping constant of man, $N \cdot sec \cdot m^{-1}$

$\bar{F}$ average vertical force during a step

k_m $m_m \omega_0^2/(1 - \zeta^2) =$ stiffness of man's muscles and reflexes acting to extend hip, knee and ankle, N/m

k_t spring stiffness of track ($= 1/\text{compliance}$), N/m

k_t^* $m_m g(l^2 - L_0^2/4)^{-1/2} =$ lowest possible track stiffness for running, N/m

L step length; distance moved during foot contact, m

L_0 step length on infinitely hard surface

m_m mass of the man, kg

m_t effective mass of the track, evaluated by Rayleigh method

t_c foot contact time on any track, sec

t_0 $\pi/\omega_0 =$ foot contact time on infinitely hard surface

u $L/t_c =$ running speed

v downward vertical velocity at moment of contact, m/sec

x_m downward displacement of the man

x_t downward displacement of the track

δ mean deflection of pillow surface in a stride, m

δ_0 shortening of the leg at mid-stance, m

ζ $b/(2\sqrt{m_m k_m}) =$ damping ratio of man

l fully extended leg length, acetabulum to heel, m

ω_n natural frequency of man and track in lowest mode of vibration, rad/sec

Subscripts

m man

o rigid-track limit

t track.

APPENDIX A

Calculation of Natural Frequency

Assume the track mass $m_t = 0$ in the schematic drawing in Fig. 3. Summing the forces acting on the track to zero,

$$(x_m - x_t)k_m + (\dot{x}_m - \dot{x}_t)b - x_t k_t = 0. \qquad (A-1)$$

Summing the forces acting on the man,

$$m_m \ddot{x}_m = -(x_m - x_t)k_m - (\dot{x}_m - \dot{x}_t)b. \qquad (A-2)$$

The frequency of the lowest mode of vibration, where the track and the man move down together, may be found by assuming a solution of the form

$$x_m = e^{i\omega t} \qquad (A-3)$$

$$x_t = A e^{i\omega t}, \qquad (A-4)$$

where A is a complex constant. Substituting equations (A-3) and (A-4) into (A-1) and (A-2)

$$(1 - A)k_m + i\omega(1 - A)b - Ak_t = 0 \qquad (A-5)$$

$$(1 - A)k_m + i\omega(1 - A)b - m_m \omega^2 = 0. \qquad (A-6)$$

Subtracting equation (A-6) from equation (A-5) gives

$$A = \frac{m_m \omega^2}{k_t}. \qquad (A-7)$$

Substituting equation (A-7) into equation (A-5),

$$\left(1 - \frac{m_m \omega^2}{k_t}\right)k_m + i\omega b\left(1 - \frac{m_m \omega^2}{k_t}\right) - m_m \omega^2 = 0. \qquad (A-8)$$

Collecting terms in ω,

$$\omega^3[i m_m b] + \omega^2[m_m(k_t + k_m)] - \omega i k_t b - k_t k_m = 0. \qquad (A-9)$$

This cubic equation was solved numerically to obtain both the real and imaginary parts of ω as a function of the parameters b, k_t, k_m and m_m. The real part of ω is called ω_n and plotted in Fig. 3 for four choices of the damping ratio $\zeta = b/(2\sqrt{m_m k_m})$.

APPENDIX B

Dimensionless Plotting

The techniques of dimensional analysis allow great simplification and reduction of labor in experimental problems where a large number of variables appear (Bridgeman, 1931). In this paper, two such problems have been discussed, the determination of step time t_c and step length L as a function of track stiffness k_t and other variables. Let us consider the dimensional analysis of each problem separately.

(a) Foot contact time

Assume that a functional relationship of the following form exists:

$$f(t_c, t_o, m_m, \zeta, k_t, k_m) = 0,$$

where

t_c = foot contact time, sec
t_o = contact time on hard surface, sec
m_m = runner's mass, kg
ζ = runner's damping ratio, dimensionless
k_t = track stiffness, N/m
k_m = man's stiffness, N/m.

One of these variables, the man's damping ratio, is already dimensionless. From the remaining five variables, the two dimensionless products t_c/t_o and k_t/k_m can be formed. A third dimensionless product using m_m, k_m and ζ may also be formed, so that the assumed form of the equation becomes:

$$\phi\left(\frac{t_c}{t_o}, \frac{k_t}{k_m}, \zeta, \frac{k_m t_o^2}{\pi^2 m_m(1 - \zeta^2)}\right) = 0.$$

The form of the last dimensionless group was chosen in such a way that its value is unity when applied to any of the

runners, according to the definition of k_m assumed in the paper. Thus the functional relationship between t_c/t_o and k_t/k_m may be determined theoretically or experimentally, and applied to any runner.

(b) Step length

In the calculation of step length, the runner's spring stiffness and damping were excluded from the problem, but his leg length and weight were assumed to be important (his weight determines the average deflection of the track surface over one stride cycle).

$$f(L, L_o, l, m_m, g, k_t) = 0,$$

where

L = step length, m
L_o = step length on hard surface, m
l = leg length, m
$m_m g$ = runner's weight, N
k_t = track stiffness, N/m.

The dimensionless form of the equation becomes

$$\phi\left(\frac{L}{L_o}, \frac{lk_t}{mg}, \frac{L_o}{l}\right) = 0.$$

The third dimensionless group, L_o/l, is assumed to be a constant for all runners. The validity of this assumption is reasonably good, as shown in Table 1.

Since k_m is assumed to be a constant, we may write the second group in the form:

$$\frac{lk_t}{mg} = \frac{k_t}{k_m}\left[\frac{(\pi/t_o)^2 l}{(1 - \zeta^2)g}\right].$$

If the term in square brackets is the same number for all runners, then a functional relationship may be found between L/L_o and k_t/k_m, as was done in Fig. 7. In fact, this term is evaluated for each of the runners in Table 1. It is not particularly constant, but is greater for the faster runners. The variation in this term explains some of the spread of the data points in Fig. 7(b) and shows why comparisons retaining the dimensions (Fig. 7a) may be preferred in this case.

(Institute for Advanced Study Mimeo, Princeton, NJ, 1935–1936); W. Pauli and V. Weisskopf, Helv. Phys. Acta **1**, 709 (1934); D. F. Moyer, Stud. Hist. Philos. Sci. **8**, 251 (1977); **9**, 35 (1978).
[22]M. J. Klein, Hist. Stud. Phys. Sci. **2**, 1 (1970); R. S. Shankland, Phys.

Rev. **49**, 8 (1936); P. A. M. Dirac, Nature **137**, 298 (1936); J. S. Jacobsen, Nature **138**, 25 (1936); N. Bohr, *ibid.* **138**, 25 (1936).
[23]P. A. M. Dirac, Proc. R. Soc. London **136**, 453 (1932); **167**, 148 (1938); **160**, 48 (1937); **165**, 199 (1938); Proc. R. Soc. Edinburgh **59**, 122 (1939).

Aerodynamic effects on discus flight

Cliff Frohlich

University of Texas, Institute for Geophysics, P.O. Box 7456, Austin, Texas 78712
(Received 30 October 1980; accepted 22 January 1981)

Skilled discus throwers claim that a properly thrown discus will travel several meters farther if it is thrown against the wind, than if it is thrown along the direction of the wind. Numerical calculations confirm these claims for winds of up to about 20 m/sec and show that the extra distance is caused by the higher lift and drag forces acting on a discus that is thrown against the wind. Aerodynamic considerations influence numerous aspects of discus throwing, but these have not been dicussed in the scientific literature. In addition to reviewing the available literature, the present article calculates the effect on distance thrown caused by changes in wind velocity, altitude, air temperature, gravity, and release velocity. Some sample results are that a discus can travel: (i) 8.2 m farther against a 10-m/sec wind than with such a wind; (ii) 0.13 m farther at 0 °C than at $+40$ °C; (iii) 0.19 m farther with no wind at the elevation of Rome, Italy than at the elevation of Mexico City, Mexico; and (v) 0.34 m farther at the equator than at the poles.

INTRODUCTION

Wind drag is an important factor affecting performance in a number of individual sports, including bicycle racing, track running (sprinting), and long jumping. Generally, for best performance it is advantageous to be moving in the same direction as the wind. In an attempt to nullify this advantage, records are disallowed in certain track and field events if there is too large a component of wind velocity along the direction of the run or jump.

Discus flight is also influenced by wind, but unlike most other track and field events, *discus throwers can throw significantly farther if the wind blows against the direction of the throw than if there is no wind or if the wind blows in the same direction as the throw.* When thrown properly, a discus is an airfoil, and the aerodynamic lift more than compensates for the loss of performance due to drag. Discus enthusiasts have been aware of this paradoxical result for many years. Almost 50 years ago Taylor[1] measured drag and lift coefficients for a discus, calculated a few trajectories, and recommended that record performances be "adjusted" for the effect of the winds. His recommendation was not instituted, and to this day discus records are allowed under any wind conditions.

Suprisingly enough, there are apparently no physics textbooks or articles in scientific journals that discuss the aerodynamics of discus flight. In fact, most of the investigations of the aerodynamics of discus flight have been reported in exceedingly obscure places. For example, the most widely quoted numerical calculation of the effect of air on discus flight is the unpublished work of Cooper *et al.*,[2] who performed their analysis as a class project for an engineering course at Purdue. Several important studies, including the best discussion on the effects of discus rotation on discus flight[3] appeared in *Discobulus*, which was a mimeographed newsletter in the 1950s for a club of British discus enthusiasts. The most comprehensive work on discus aerodynamics is available only in Russian.[4] Measurements of the drag and lift coefficients for a discus at various angles of attack have been published in a physical education journal by Ganslen.[5] Several other authors have presented summaries of portions of the above work, including Lockwood,[6] Dyson,[7] and Hay.[8]

Many basic questions about the effect of physical variables on discus flight are not addressed at all by any of the previous work. As an airfoil, how will the discus be affected by changes in air density, Earth gravity, and its own mass and shape? Will a discus perform better when thrown at high altitudes and high air temperatures or at low altitudes and colder air temperatures? How much further will a discus travel if thrown at the Earth's poles than at its equator? If they are released at the same velocity, which will travel further, a men's discus or the smaller and lighter women's discus (see Table I).

FACTORS INFLUENCING DISCUS FLIGHT
Definitions of basic variables

When in flight, a discus is affected only by the forces of gravity, aerodynamic drag, and aerodynamic lift (see Fig. 1). If there is wind with nonzero velocity $\mathbf{v}_w$ then the aerodynamic drag will not act along a direction opposing the velocity $\mathbf{v}_d$ of the discus, but rather it will act along the direction of the relative velocity $\mathbf{v}_{\text{rel}}$ (see Fig. 1) where

$$\mathbf{v}_{\text{rel}} = \mathbf{v}_d - \mathbf{v}_w. \tag{1}$$

The magnitude of the drag and lift forces are usually represented in terms of the dimensionless drag and lift coefficients c_d and c_L:

$$F_{\text{drag}} = \tfrac{1}{2}c_d\rho A v_{\text{rel}}^2; \quad F_{\text{lift}} = \tfrac{1}{2}c_L\rho A v_{\text{rel}}^2, \tag{2}$$

where ρ is the density of the air and A is the maximum

Table I. Characteristics of the discus and frisbee.

	Mass (kg)	Diameter (mm)	Thickness (mm)	M_{eff} (kg)	Current world record [d] (m)
Men's discus [a]	2.0	221	46	2.0	71.16
High school discus [b]	1.616	211	41.28	1.77	63.86
Women's discus [a]	1.0	182	39	1.47	70.72
Wham-O regular frisbee [c]	0.087	227	31	0.08	

[a] Reference 9.
[b] Reference 10.
[c] S. Johnson, *Frisbee* (Workman, New York, 1975).
[d] B. Nelson, Track & Field News **32**(12), 16 (1980).

cross-sectional area of the discus. From Eq. (2), the acceleration due to aerodynamic forces is

$$a = \tfrac{1}{2}\rho A v_{rel}^2 (c_d^2 + c_L^2)^{1/2}/M. \qquad (3)$$

c_d and c_L depend strongly on the attack angle ψ, which is the angle between the plane of the discus and the direction of $\mathbf{v}_{rel}$. They may also depend weakly on $\mathbf{v}_{rel}$, ρ, and the discus rotational velocity ω. For simplicity, in the numerical calculations presented in this paper, it has been assumed that c_d and c_L are independent of v_{rel}, ρ, and ω, and that the rotation vector ω is perpendicular to the plane of the discus.

Equations of motion

The equations of motion are particularly simple if $\mathbf{v}_w$, $\mathbf{v}_d$, and the normal to the plane of the discus all lie within a vertical plane, i.e., if the discus is thrown either with or against the wind, and so that the discus does not lean to the right or the left. In this case,

$$\ddot{x} = -\tfrac{1}{2}(\rho A v_{rel}^2/M)(c_d \cos\beta + c_L \sin\beta),$$
$$\ddot{y} = -g + \tfrac{1}{2}(\rho A v_{rel}^2/M)(c_L \cos\beta - c_d \sin\beta), \qquad (4)$$

where β is the angle between the horizontal plane and v_{rel}. In addition if the lift and drag forces apply no torques to the discus, then the angle α between the plane of the discus and the horizontal remains constant throughout the flight. In this case, the path of the discus is determined completely by Eq. (4) if the initial conditions are known. The initial conditions include the initial release velocity v_{d_0}, the wind velocity v_w, the release angle R (the angle between the horizontal plane and the initial velocity vector), the discus inclination angle α (see Fig. 1), and the release height y_0.

These equations can be solved if the initial conditions are known. In particular, at time T after the discus has been released with initial velocity

$$\mathbf{v}_{rel_0} = \mathbf{v}_{d_0} - \mathbf{v}_w:$$

$$\dot{x} - v_w = v_{rel_x} = v_{rel_{x_o}} - \tfrac{1}{2}(\rho A/M)$$
$$\times \int_0^T v_{rel}^2 (c_d \cos\beta + c_L \sin\beta)dt',$$

$$\dot{y} = v_{rel_y} = -gT + v_{rel_{y_o}} + \tfrac{1}{2}(\rho A/M)$$
$$\times \int_0^T v_{rel}^2 (c_L \cos\beta - c_d \sin\beta)dt'.$$

Thus

$$x = v_w T + v_{rel_{x_o}}T - \tfrac{1}{2}(\rho A/M)$$
$$\times \int_0^T \int_0^t v_{rel}^2 (c_d \cos\beta + c_L \sin\beta)dt'dt, \qquad (5)$$

$$y = -\tfrac{1}{2}gT^2 + v_{rel_{y_o}}T + \tfrac{1}{2}(\rho A/M)$$
$$\times \int_0^T \int_0^t v_{rel}^2 (c_L \cos\beta - c_d \sin\beta)dt'dt.$$

In this form, the solution neatly separates the effects of gravity g and wind v_w from the aerodynamic forces depending on v_{rel}, but it is not particularly useful for numerical calculations.

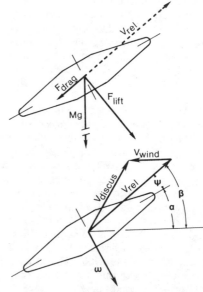

Fig. 1. Top: forces affecting discus flight include gravitational forces and aerodynamic forces. By definition the components of aerodynamic forces along the direction of the oncoming air are called drag forces, and the components perpendicular to the direction of the oncoming air are called lift forces. In this figure, the velocity of the discus relative to the air is such that the air strikes the upper surface of the discus, and so the direction of the lift is downward. Bottom: important variables controlling aerodynamic forces on the discus. The relative velocity $\mathbf{v}_{rel}$ depends on the actual discus velocity $\mathbf{v}_{discus}$ and the wind velocity $\mathbf{v}_{wind}$. The drag and lift coefficients depend strongly on the attack angle ψ. If the aerodynamic forces apply no torques to the discus, then the angle α between the plane of the discus and the horizontal does not change during flight. If torques do exist, their effect on the discus depends on its angular velocity.

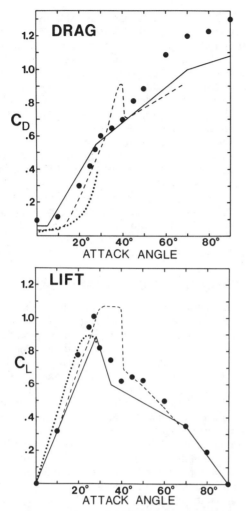

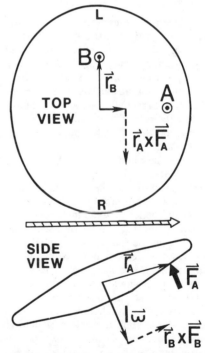

tests showed that accurate results could be obtained with time increments of 0.1 sec. With the 0.1-sec increments, calculated distances for typical throws varied less than 0.1 m from distances calculated using an increment of 0.01 sec.

Effect of rotation of the discus

The most important effect of the discus' rotation is to stabilize its orientation during flight. Upon release the discus rotates at about 7 rps.[1] In the absence of applied torques the discus' initial orientation is preserved throughout its flight.

At the moment of release, the discus thrower will usually attempt to orient the discus so as to maximize the lift forces and minimize the drag forces while the discus is traveling upward and outward. Most investigators agree that the optimum strategy in still air is to release the discus so that its inclination angle α is about 5° to 10° less than the release angle R.[2,5] Although this results in negative lift during the very beginning of flight, it allows for a minimum of drag and optimum average lift throughout the upward part of its flight.

In fact, the torques applied by aerodynamic forces during flight are not negligible, although they are quite small. If one carefully observes the orientation of a discus in flight, one notices that for a right-handed thrower the left side of the discus tilts (rolls) gradually downward about 10° during the latter portion of its flight.[3] As explained in Fig. 3, this is because the aerodynamic lift forces F_A are appar-

Fig. 2. Drag and lift coefficients for the discus. Solid circles are the measurements of Ganslen[5,8] for a relative air velocity of 24.4 m/sec. The solid line represents the values used in the numerical calculations of Cooper *et al.*[2] and in this paper. The dotted lines are the results of Kentzer and Hromas[3] for a relative velocity of 30.5 m/sec and a rotation of 2.5 rev/sec. The dashed lines are the coefficients reported by Tutevich.[4]

In practice, it is easier to calculate discus trajectories by making use of Eq. (4) and integrating over small time increments Δt. Since the drag and lift coefficients are nearly piecewise linear with the attack angle ψ (Fig. 2), numerical integration is improved by making use of $\ddot{x}$ and $\ddot{y}$. Since

$$\beta = \tan^{-1}(v_{\mathrm{rel}_y}/v_{\mathrm{rel}_x}),$$

$$\dddot{x} = (\rho A/2M)\{2(v_{\mathrm{rel}_x}\ddot{x} + v_{\mathrm{rel}_y}\ddot{y})(-c_d\cos\beta - c_L\sin\beta)$$
$$+ (v_{\mathrm{rel}_x}\ddot{y} - v_{\mathrm{rel}_y}\ddot{x})[(\partial C_d/\partial\psi - c_L)\cos\beta$$
$$+ (\partial C_L/\partial\psi - c_d)\sin\beta]\},$$

$$\dddot{y} = (\rho A/2M)\{2(v_{\mathrm{rel}_x}\ddot{x} + v_{\mathrm{rel}_y}\ddot{y})(c_L\cos\beta - c_d\sin\beta)$$
$$+ (v_{\mathrm{rel}_x}\ddot{y} - v_{\mathrm{rel}_y}\ddot{x})[(-\partial C_L/\partial\psi - c_d)\cos\beta$$
$$+ (\partial C_d/\partial\psi - c_L)\sin\beta]\}.$$

Thus over each time increment Δt,

$$\Delta x = \dot{x}\Delta t + \tfrac{1}{2}\ddot{x}\Delta t^2 + \tfrac{1}{6}\dddot{x}\Delta t^3,$$

$$\Delta y = \dot{y}\Delta t + \tfrac{1}{2}\ddot{y}\Delta t^2 + \tfrac{1}{6}\dddot{y}\Delta t^3.$$

The author has written a FORTRAN program using this method to calculate trajectories for the discus, and

Fig. 3. One effect of air on the discus is to create torques that change the orientation of the axis of rotation of a discus. For a discus tilted with front slightly upwards and moving to the right the largest torques are caused because the lift forces are larger on the forward half of the discus than the rear half. Thus the lift force F_A acts at point A and creates a torque $\mathbf{r}_A \times \mathbf{F}_A$ that causes the angular momentum vector $I\omega$ to move in space. This causes the right edge of the discus to tilt slowly upwards during flight. Similarly, because of the discus rotation, the relative velocity is greater on the left side of the discus than the right, causing a net upwards force F_B to act on the left side of the discus. This force creates a torque $\mathbf{r}_B \times \mathbf{F}_B$, causing the forward edge of the discus to tilt progressively upwards during flight.

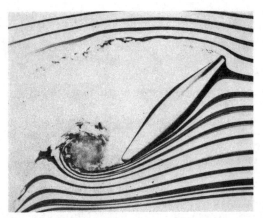

Fig. 4. Wind tunnel experiments demonstrate that the abrupt decrease in lift that occurs as the attack angle approaches 30° is accompanied by the formation of a turbulent wake. This wake is clearly visible in this wind tunnel photo of a discus at an attack angle ψ of 45°. Smoke has been introduced into the airstream so that the streamlines can be followed. This photo is reproduced with the permission of Richard Ganslen, who produced it when he was at Purdue University.

ently larger on the forward half of the discus, creating a torque vector pointing to the right. Since the angular momentum vector points mostly downward, and since torque equals the rate of change of angular momentum, the torque to the right causes the left side of the discus to tilt downwards.

Similarly, since the discus' rotation causes the relative air velocity to be slightly higher on the left side of the discus, the aerodynamic forces effectively are applied slightly to the left of the center of mass (Fig. 3). These forces create a torque vector pointing forward, causing the front edge of the discus to tilt (pitch) upward about $1\frac{1}{2}$°/sec during the course of its flight. Presumably aerodynamic forces during flight also slow the rate of rotation of the discus about its own axis, however, this effect has not been measured and cannot be very large.

Although the mass and shape of the discus are quite strictly controlled by the rules governing the sport,[9,10] there are no restrictions on the moment of inertia or the distribution of mass within the discus. Recently, most discus throwers have tended to prefer implements where most of the mass is concentrated near the rim. Such a discus has a large moment of inertia that resists changes in the discus orientation caused by aerodynamic torques. Of course, because of its symmetry, a nonrotating discus would also experience smaller aerodynamic torques than a rotating discus. This has caused at least one expert to suggest that a discus be constructed with a hollow mercury-filled rim in order to reduce the rotation after release (Ruudi Toomsalu, personal communication). However, experience suggests that a nonrotating discus lacks stability and will wobble and flutter, and thus the liabilities of a fluid-rim discus seem to be outweighed by the benefits.

Review of experimental measurements of c_d, c_L, and pitching moments

All reported investigations of the drag and lift coefficients for the discus agree that the drag coefficient increases steadily as the attack angle ψ increases from 0° to 90° (see Fig. 2).[2–5] In addition, they find that the lift is zero when $\psi = 0$°, it increases quickly to a maximum value at

about $\psi = 30$°, and then drops quite sharply and approaches 0 as ψ approaches 90°. By studying the behavior of smoke streams passing a discus suspended in a wind tunnel (Fig. 4), Ganslen[5] was able to show that the abrupt decrease in lift at about 30° coincides with the formation of a turbulent wake behind the discus. Of the reported measurements of c_d and c_L, those of Ganslen[5] are the most reliable, and he is the only investigator reporting measurements of drag and lift coefficients over a range of air velocities. The basic form of the coefficients does not vary as the velocity changes, although the value of the drag and lift coefficients decreased by about 30% as the velocity of the wind increased from 21 to 30 m/sec.

The only published measurements of rotational torques on a rotating discus are those of Kentzer and Hromas,[3] who reported only the pitching moments. They found that pitching moment varies in a fashion similar to the lifting force, which apparently provides the force causing the moment.

RESULTS OF NUMERICAL CALCULATIONS

If reliable measurements existed at all attack angles for drag and lift coefficients, pitching and rolling moments, and discus rotation slowing torques, it would be possible to write a computer program to calculate the discus trajectory if eight initial conditions were known: the discus initial rotation speed and orientation vector (three variables), the initial release velocity (two variables if the azimuth of the throw is along the x axis), the release height (one variable) and the wind velocity (two variables, if horizontal winds only are assumed). In practice it is not practical to search for optimum combinations of eight variables, especially since the rotation moments have not been measured. For this reason, a number of simplifying assumptions will be made in order to simplify the calculations.

In addition, in all the calculations it has been assumed that wind velocity is independent of time and of height above the ground, and that the discus rotates only about an axis perpendicular to its faces (no "wobble"). Except as noted, the drag and lift coefficients of Cooper et al.[2] are used (see Table II) since they are the simplest and exhibit the same basic form as the other measurements. It will be assumed that pitching, rolling, and rotation slowing moments are zero, that the wind direction is parallel to the direction of the throw, and that the plane of the discus intersects the plane of the ground along a line perpendicular to the direction of the throw (i.e., the discus tilts neither to the right or the left). With these assumptions, Eq. (4) applies, and only five variables remain, R, α, v_{d_0}, v_w, and y_0.

Aerodynamic forces are definitely an important factor in discus flight. For example, the ratio a/g of aerodynamic to

Table II. Drag and lift coefficients of Cooper et al. (Ref. 2) used for numerical calculations in the present paper. Linear interpolation may be used to find coefficients between table entries.

	Drag			Lift
Angle	C_d		Angle	C_L
0°	0.06		0°	0.00
5°	0.06		28°	0.875
30°	0.54		35°	0.60
70°	1.00		70°	0.35
90°	1.07		90°	0.00

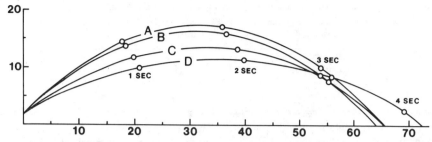

Fig. 5. Trajectory of a discus thrown with optimum release angle R and inclination angle α under various wind conditions, including: A—no air, discus thrown in a vacuum; B—a 10-m/sec wind in the direction of the throw; C—no wind; and D—a 10-m/sec wind against the thrower. Open circles show the point on the trajectory at each second after the time of release. Note that at these wind velocities, the longest distances are obtained when throwing against the wind with a low, flat trajectory. The release velocity is 25 m/sec, $A = 0.038$ m^2, $M = 2.0$ kg, $\rho = 1.29$ kg/m^3, $g = 9.8$ m/sec^2, release height $y_0 = 1.8$ m, and the drag and lift coefficients are those of Cooper et al.[2]

gravitational forces during discus flight is [see Eq. (3)]:

$$a/g = (\rho A v_{\rm rel}^2 / 2Mg)(c_d^2 + c_L^2)^{1/2}. \qquad (6)$$

Typically, this ratio will approach or exceed unity during some part of the discus flight. In this section we will consider the effects on distance thrown of variations in each of the parameters which contribute to this ratio.

Effect of v_w

Champion discus throwers claim that longer throws can be made throwing against fairly stiff winds than with the wind or with no wind[11] and my calculations confirm this result (see Fig. 5). For moderate winds, the worst possible situation is to throw with a wind whose velocity is 7.5 m/sec (Fig. 6). A discus will travel more than 6 m further if thrown against a 7.5-m/sec wind than if thrown with the wind. For winds to about 20 m/sec, a properly thrown discus will always travel further against the wind than with it. Thus under all conditions under which a discus competition could conceivably be held, discus throwers desiring record performances are correct in their preference for throwing in the face of stiff wind.

However, it is clear that this cannot be true for very high winds. This can be shown with an approximate calculation. Supposing the discus is thrown with an attack angle of zero (so that the drag is a minimum) and suppose it is released with a release angle of zero (so that its rate of horizontal progress is at a maximum). If the wind speed is high enough, the discus will stop its forward progress and travel back towards the thrower before it has traveled 60 m. In particular, if the attack angle and the release angle are zero, the horizontal acceleration against the direction of the throw is then at least [see Eq. (4)]:

$$\frac{dv_d}{dt} = -\frac{1}{2}\frac{\rho A c_d}{M}(v_d - v_w)^2.$$

This is integrable, with

$$v_d = v_w - 1/(K - Xt),$$

where K is a constant of integration that equals $-1/(v_{d_0} - v_w)$ if the initial horizontal velocity of the discus is v_{d_0}, $X = \frac{1}{2}\rho A c_d / M$ and t is time. The wind will decelerate the discus and when $v_d = 0$ the discus will reach its maximum distance $X_{\max}$. A straightforward calculation reveals that

$$X_{\max} = \frac{1}{X}\left(\frac{-1}{1 - v_w/v_{d_0}} + \ln\left|1 - \frac{v_{d_0}}{v_w}\right|\right).$$

Of course, this is an overestimate of the distance thrown, since the assumed drag force was the minimum possible force. However, for $c_d = 0.06$, $v_w = 3v_{d_0}$, and ρ, A, and M as in Fig. 5, one finds that $X_{\max}$ is 51 m. This confirms that optimum throws cannot be made against hurricane force winds.

For moderate winds (less than about 20 m/sec) a discus thrown with the wind must be thrown with a different strategy than a discus thrown against the wind. Against the wind, the discus inclination α at the moment of release should be about 10° to 15° less than the release angle R, so that during most of the flight the drag will be a minimum and the lift a maximum. As the wind velocity increases, the release angle R decreases, so that the trajectory becomes flatter, and the discus will not hang in the air too long and "catch" the wind at the end of the throw. In contrast, a discus thrown with the wind (Fig. 5) should be lofted higher in the air (R is larger), and in fact as the wind velocity reaches about 20 m/sec longer throws will be obtained if the discus is turned over so that the face of the discus catches the wind like a sail.

Although the longest throws occur against stiff winds, it is easier to obtain near-optimum performances when throwing with the wind. To obtain a throw within 1 m of the optimum when throwing against a 10-m/sec wind, the

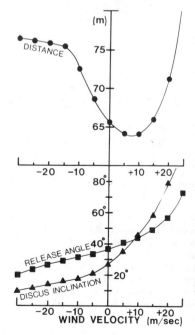

Fig. 6. Top: longest possible throw at various wind velocities. The solid line is the distance interpolated between the calculated points (solid circles). Note that the worst possible conditions to obtain long throws is to throw with a wind of 7.5 m/sec, and that if the wind velocity is less than about 20 m/sec, longer throws can always be obtained by throwing against the wind (negative wind velocities). ρ, g, A, M, y_0, c_d and c_L as in Fig. 5. Bottom: the release angle R (squares) and discus inclination α (triangles) that produce the longest throws at each wind velocity.

Table III. Approximate effect of variations in environmental conditions on maximum distance thrown with a men's discus released at 25 m/sec.

	Variation	0-m/sec wind	− 10-m/sec wind
Temperature	+ 10°C	− 3.4 cm	− 22.6 cm
Atmospheric pressure	+ 10 mm Hg	+ 1.2 cm	+ 8.1 cm
Elevation	+ 1 km	− 8.5 cm	− 55.8 cm
Gravity	+ 1 cm/sec²	− 6.5 cm	− 7.5 cm

athlete must control the release angle R to within about ± 5° and the discus inclination α to within about ± 3° (Fig. 7). However, when throwing with a 10-m/sec wind, he can come within 1 m of the optimum by controlling these angles to within only ± 6° and ± 15°. In competition, throwing against the wind would seem to favor experienced discus throwers, since more control is needed to obtain optimum performances.

The measured drag and lift coefficients are probably not accurate for relative velocities above about 40 m/sec, and so it is not meaningful to perform trajectory calculations for winds with velocities above about 20 m/sec. However, it is interesting to note that for optimum throws both with and against the wind at 20 m/sec the discus inclination is nearly parallel to the relative velocity vector at the point on the trajectory when the relative velocity (and thus drag) is a maximum. For throws against the wind, this occurs at the moment of release, and thereafter drag reduces the relative velocity. For throws with the wind, this occurs at the end of the discus flight. The following wind lifts the discus and increases its potential energy, and then the relative speed increases as the discus falls and loses potential energy.

Effect of ρ, A, and M

From Eq. (3), the effect of aerodynamic forces is proportional to the quantity $\rho A / M$. The discus area A and mass M

are fixed by the rules governing the sport (Table I), but the air density ρ can vary nearly 50% between a high-temperature, high-altitude site and a low-temperature, low-altitude site (Table III). Although an athlete can do nothing to vary ρ to his favor during a particular competition, he may select competitions with favorable altitude and temperature if he desires to throw record distances.

A convenient quantity for calculating the effect of changing ρ, A, and M on distance is the effective mass, defined by

$$M_{\mathrm{eff}} = M\,(\rho_{\mathrm{STP}}/\rho)(A_{\mathrm{men's}}/A\,), \qquad (7)$$

where $\rho_{\mathrm{STP}} = 1.29$ kg/m³ and $A_{\mathrm{men's}} = 0.038$ m². From Eq. (4), it is clear that for given initial release conditions two discuses with the same effective mass will possess similar trajectories. Thus the effective mass not only allows a simple comparison of the effect of changing air density, but also allows us to compare trajectories for implements with different surface area A and mass M.

The calculations show that changes in the effective mass do affect the distance of the throw (Fig. 8). The increase in distance due to decreasing the effective mass is not highly significant when there is no wind, but is about six times larger when the discus is thrown into a 10-m/sec wind. Figure 8 and Table III show that with no wind, changes in altitude and temperature can affect distance thrown by less than a meter, but with a wind the effect can be as large as several meters. Similarly, a women's discus will travel farther than a men's discus if both have the same release velocity. Both men and women attempting to obtain record per-

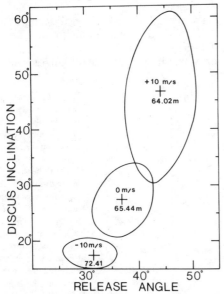

Fig. 7. Effect of changes in release angle and angle of inclination on distance thrown at various wind velocities. Plotted points (+) show optimum release angle and discus inclination for winds of 10 m/sec for and against the thrower, and for no wind. The solid line surrounding the optimum throw outlines the region containing release angle and discus inclinations that result in a throw within one meter of the optimum distance. Note that it is much easier to produce near optimum throws when throwing with the wind. ρ, g, A, M, y_0, c_d, and c_L are as in Fig. 5.

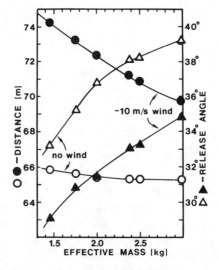

Fig. 8. Calculated optimum distance thrown (circles) and optimum release angle (triangles) versus effective mass (defined in text) for no wind (open symbols) and a 10-m/sec wind against the thrower (filled symbols). A discus with smaller effective mass is influenced more by aerodynamic forces than a discus with larger effective mass, and thus will travel somewhat further. For these calculations $v_{d_0} = 25$ m/sec, and g, y_0, c_d, and c_L are as in Fig. 5.

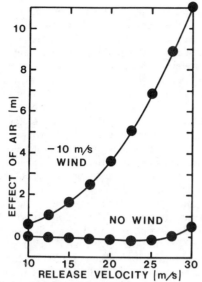

Fig. 9. Difference between optimum distance thrown in air and in a vacuum. In a vacuum, the optimum distance D is given by the equation $D = (v_{d_0}^2/g)(1 + 2gh/v_{d_0}^2)^{1/2}$ (see Lichtenburg and Wills[13]). Distances in air are calculated assuming ρ, A, y_0, M, c_d, and c_L as in Fig. 5.

formances should throw at low altitudes and low temperatures.

From the athlete's standpoint, the most significant result of decreasing the effective mass is that the release angle must be decreased in order to obtain the optimum distance. Thus for optimum performance women discus throwers should release the discus at an angle several degrees smaller than indicated in Fig. 7, and men throwing at high temperatures and altitudes should release it at an angle several degrees larger than indicated in Fig. 7.

Effect of v_{d_0}

To improve performance the single most important variable that a discus thrower can affect by training is his release velocity v_{d_0}. It is for this reason that training for discus throwers usually consists primarily of weight lifting to increase strength and agility drills to improve body quickness.[12] It is also for this reason that tall, strong athletes with long arms are most likely to be successful as discus throwers.

Aerodynamic forces become more important as the release velocity increases (Fig. 9). Up to about 25 m/sec with no wind, the optimum distance thrown is nearly the same whether the discus is thrown in a vacuum or in air, although the angle of release that should be used to obtain the optimum throw is quite different (see Lichtenberg and Wills[13] for a discussion of optimum release angles for projectiles in a vacuum). One can actually throw a discus further in the presence of air than in a vacuum if the release velocity is higher than 25 m/sec, or if one is throwing against a stiff wind (Fig. 9).

Other factors: effect of gravity g, release height y_0, and drag and lift coefficients c_d and c_L

In practice, neither gravity nor release height affect discus performance as significantly as the factors discussed above. Variations in g have little effect on discus flight, because g varies less than about 0.5% (or about 5 cm/sec^2)

over the surface of the Earth. An increase in gravity of this magnitude may be expected to reduce the distance that a discus is thrown less than 0.34 m (see Fig. 10 and Table III).

The author has performed no explicit calculations concerning the effect of changing the release height y_0. A crude estimate of the effect of changing y_0 is that it would be proportional to cot θ, where θ is the angle between the discus trajectory and the Earth's surface at the point the discus strikes the ground, i.e.,

$$\Delta_{\text{distance}} = \Delta y_0 \cot \theta.$$

From Fig. 4 it appears that a change in y_0 of 1 m would change the distance thrown by about 2 m. However, in practice it is very unlikely that a discus thrower could maintain top form and change y_0 by more than a small fraction of a meter.

As reported earlier (Fig. 2), the measured drag and lift coefficients all have nearly the same general form, although at a given angle of attack the actual values vary somewhat. Since c_d and c_L only affect the aerodynamic forces on a discus, it is clear from Eqs. (3) and (4) and Fig. 8 that a proportional increase in drag and lift coefficients would have exactly the same effect on discus flight as a proportional increase in air density or a proportional decrease in the effective mass. Although a proportional increase in c_d and c_L will increase the distance thrown against winds of 10 m/sec (Fig. 8), increasing c_d while holding c_L constant will actually decrease the optimum distance thrown. In fact, the author's calculations found that doubling c_d at all angles of attack decreased the optimum distance thrown by 15.8 m, and halving c_d increased the optimum distance by 19.4 m.

DISCUSSION AND CONCLUSIONS

Aerodynamic forces are significant factor in discus flight, approaching the magnitude of gravitational forces under certain conditions. This allows a trained thrower to throw a discus several meters further against a wind than would be possible if no aerodynamic forces or no wind were present. It is because of the importance of aerodynamic factors that the discus must be thrown with a release angle of about 30° to 40° instead of the textbook value of about 45°. In addition, for maximum distance it must be oriented so as to take maximum advantage of the effects of drag and

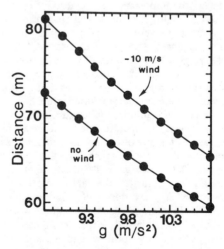

Fig. 10. Effect of changes in the Earth's gravity on the optimum distance thrown. ρ, A, y_0, M, c_d, and c_L are as in Fig. 5.

Fig. 11. Wham-O regular frisbee, and a Gill High School wood center discus. Because a discus and a frisbee are of similar size and shape, but different mass, they experience similar aerodynamic forces but vastly different gravitational forces.

lift forces during different portions of the flight.

As an airfoil, the discus is remarkably similar to its less massive cousin the frisbee (see Fig. 11 and Table I). In fact, in high winds a discus must be thrown just like a frisbee would be thrown under normal conditions, with a low release angle and a flat inclination. This is because the ratio of aerodynamic to gravitational forces is of similar magnitude. In particular, for a frisbee when $\psi = 10°$ and if $v_{d_0} = 5$ m/sec,

$$a/g - 0.184,$$

and for a discus if $v_{d_0} = 30$ m/sec, as might be encountered when throwing into a moderate wind,

$$a/g = 0.274.$$

Apparently the mass and shape of the discus place it in a regime where increasing the aerodynamic forces on it will increase the distance thrown. In particular, the numerical calculations presented in this paper indicate that for a given release velocity longer distance will be obtained with higher air densities, lower temperatures, lower altitudes, lighter discuses, larger discus areas, and stronger winds against the direction of the thrower. However, this trend cannot continue for extreme aerodynamic conditions, as indicated by the calculation showing that a discus must travel less far when thrown against exceptionally high winds. Experience suggests that a frisbee is probably in a regime where increasing the aerodynamic forces on it will decrease the distance thrown.

Considering the importance of the aerodynamic forces affecting discus flight, it is surprising that record books make no distinctions between marks recorded in still air and wind-aided marks. Apparently the present paper is the first work in English to quantitatively estimate the effect on distance thrown caused by changes in temperature, altitude, discus size, and gravity. It appears that a 10-m/sec wind is a more important factor in discus than in sprints,[14] the long jump or triple jump, or in the shot put.[13] Because they affect air density, temperature and altitude are also an important factor affecting discus results. Differences in air density alone would cause a discus released at 25 m/sec on a cold day (− 10 °C) in Moscow (elevation 120 m) to travel approximately 0.36 m further than an identical discus released in Mexico City (elevation 2239 m) in warm temperatures (+ 40 °C). Similarly, because it is lighter, a women's discus will experience relatively larger aerodynamic forces than a men's discus under otherwise identical conditions and thus can be thrown further if it has the same release velocity.

It is possible that head winds possessing a component blowing from the (right-handed) discus thrower's right side would permit even longer throws than those obtained with direct headwinds. This would explain why many champion throwers prefer to throw in "quartering winds." Under these conditions, the thrower would want to orient the discus with its highest point slightly to the right of straight ahead. Because the relative velocity would be smaller, both drag and lift would be reduced, but longer distances might be obtained since some of the drag forces would act perpendicular to the direction of flight. The question of whether longer distances would actually be attained has not been attacked quantitatively in the present study, but is the focus of an ongoing study.

Perhaps the most outstanding problem that remains to be studied quantitatively is the effect of rotation of the discus on its flight. It is possible to throw a discus in the wind in such a fashion that the aerodynamic torques cause it to pitch forward during flight, thereby allowing the discus to maintain a near-optimum angle of attack throughout a greater portion of the entire trajectory? Although it is unlikely that this would improve performance, it would be relatively easy to extend the author's numerical calculations to include the effects of rolling and pitching torques on the discus. However, no reliable wind tunnel measurements of these torques are available, and apparently no wind tunnel measurements of drag and lift coefficients have been performed in the last decade.

ACKNOWLEDGMENTS

Several individuals were kind enough to review an earlier draft of this manuscript, including David Caughy of Cornell University, Richard Ganslen of Denton, Texas, and Doug McCowan of the University of Texas. Barry Willis, of Great Britain, Ruudi Toomsalu of Estonia, and Richard Ganslen provided source materials that were unavailable in libraries in the United States. I am also grateful for conversations with Bob Cowles of Cornell University. This work has the University of Texas, Institute for Geophysics contribution number 458.

[1]J. A. Taylor, Athletic J. **12**, 9 (1932).
[2]L. Cooper, D. Dalzell, and E. Silverman, Flight of the Discus (unpublished, 1959). Available in 1980 from Purdue University Library.
[3]C. P. Kentzer and L. A. Hromas, Discobulus **4**, 1 (1958). Available in 1979 from Barry Willis, 14 Blueridge Avenue, Brookmans Park, Hatfield, Herts, Great Britain.
[4]V. N. Tutevich, *Teoria Sportivnykh Metanii* (in Russian, Moscow, 1969).
[5]R. V. Ganslen, Athletic J. **44**, 50 (1964).
[6]H. H. Lockwood, in *Athletics*, edited by G. F. D. Pierson (Nelson, Edinburgh, 1963).
[7]G. Dyson, *The Mechanics of Athletics*, 7th ed. (Holmes and Meier, New York, 1977).
[8]J. G. Hay, *The Biomechanics of Sports Techniques* (Prentice-Hall, Englewood Cliffs, NJ, 1978).
[9]H. Rico and J. Jackson, *Official AAU Track and Field Rules 1977* (Amateur Athletic Union of the U.S., Indianapolis, IN, 1977).
[10]*The Official Track and Field Guide 1979* (National Collegiate Athletic Association, Shawnee Mission, KS, 1978).
[11]K. Stone, Track & Field News **30**(6), 10 (1978).
[12]W. Paish, *Discus Throwing*, 4th ed. (British Amateur Athletic Board, London, 1976).
[13]D. B. Lichtenburg and J. G. Wills, Am. J. Phys. **46**, 546 (1978).
[14]C. R. Kyle, Ergonomics **22**, 387 (1979); L. Pugh, J. Physiol. **213**, 255 (1971).

J. Biomechanics Vol. 18, No. 5, pp. 337–349, 1985.
Printed in Great Britain

0021–9290/85 $3.00 + .00
Pergamon Press Ltd.

A MATHEMATICAL THEORY OF RUNNING, BASED ON THE FIRST LAW OF THERMODYNAMICS, AND ITS APPLICATION TO THE PERFORMANCE OF WORLD-CLASS ATHLETES

A. J. WARD-SMITH

Department of Mechanical Engineering, Brunel University, Uxbridge, UB8 3PH, U.K.

Abstract—Following a survey of existing mathematical models of running, a new analysis is developed, based on the first law of thermodynamics. The method properly accounts for each term in the energy balance, and avoids the use of mechanical efficiency factors. A relationship is derived between race distance and the time taken to run that distance. An excellent correlation of results from recent Olympic Games is established for events over distances from 100 m to 10,000 m. The velocity–time relationship for a sprinter running 100 m at maximum available power is obtained by numerical integration of the power equation. It is shown that the peak velocity is achieved in the middle stages of the race, a result which is consistent with practice, but which previous calculations based on Newton's laws have failed to predict. Further applications of the analysis are indicated.

NOMENCLATURE

A	rate of degradation of mechanical energy into thermal energy, per unit velocity
a	constant coefficient, defined by equation (2)
b	constant coefficient, defined by equation (2)
C	chemical energy released
C_D	drag coefficient
D	aerodynamic drag
E	energy available from the anaerobic mechanisms at time t
E_o	capacity of the anaerobic mechanisms
E_a	capacity of the alactacid mechanism
E_g	capacity of the glycolitic mechanism
E_R	energy required for running
F_D	driving force
F_R	resistive force
g	acceleration due to gravity
H	mechanical energy degraded into thermal energy
K^*	aerodynamic parameter defined by equation (29)
m	mass of the runner
P	power available from the conversion of chemical energy
P_e	external power
P_H	rate of degradation of mechanical energy into thermal energy
P_{max}	maximum power available from chemical energy conversion
R	steady rate of energy release, measured above rate at rest
S	frontal projected area of runner
s	total distance of event
S_o	parameter defined by equation (23)
T	time taken to run distance s
t	time from rest
v	instantaneous velocity
V_{max}	maximum velocity
W	external work done by centre of mass of runner
x	horizontal distance traversed from rest
ρ	air density
λ	parameter governing the variation of power with time

Superscript

*	denotes a quantity divided by m (e.g. $C^* = C/m$)

1. INTRODUCTION

Over the past 60 yr there has been a consistent interest in the mathematical analysis of running performance and, in particular, in prediction methods which relate distance and running time. Lloyd (1966, 1967), using a method based on energy considerations, established an analysis applicable to both sprinting and distance events. Margaria (1976) also refers briefly to running performance in long distance races. However, ever since the publication of the paper by Furusawa *et al.* (1927a), attention has been focussed primarily on the sprinting events, and the principal vehicle of analysis has been the application of Newton's second law of motion. A survey of existing calculation methods is made in section 2 of the present paper.

The purpose of the present paper is to derive a new analysis of running, based on energy considerations, and at this stage it may be helpful to establish the perspective of the present analysis.

During every stride of running the human body is subject to a whole range of energy exchanges (see, e.g. Ward-Smith, 1984), of which the principal components are as follows: muscles generate mechanical energy by the conversion of chemical energy, the levels of kinetic energy and potential energy of the limbs continually change, the body gains and loses kinetic and potential energy as it moves vertically in the earth's gravitational field, strain energy is stored and released in muscles and tendons, external work is done against aerodynamic drag, kinetic energy is added to the runner's centre of mass by horizontal acceleration, and a considerable amount of energy is degraded into thermal energy. Each of these components must be

Received 6 December 1983; in revised form 1 August 1984.

carefully appraised when energy balances are considered over the course of a single stride.

During the course of a race an athlete makes a large number of strides, and it is appropriate to consider the energy charges that occur over the course of the race relative to the resting state at the start of the race. The components considered above again feature in the energy balance, but it is convenient to divide the elements into two separate categories. The first category contains those components whose maximum values are limited by the cyclical nature of the stride pattern, and includes the kinetic energy and potential energy of the limbs, the kinetic energy and potential energy associated with the vertical movement of the body in the earth's gravitational field, and the strain energy stored within the muscles and tendons. The second category contains those components whose magnitude increases inexorably as the number of strides increases, and includes the amount of chemical energy transformed into mechanical energy, the external work done against aerodynamic drag and the mechanical and chemical energy transformed into thermal energy. It is also convenient to include in this second category the kinetic energy added to the centre of mass of the runner, which increases from rest in the early phase of a race, but, depending on the type of race, may diminish slightly over the later stages.

As the race proceeds the magnitudes of the energy terms in the second category become progressively larger and, in computing energy balances, the detailed contributions of the individual terms in the first category become less significant on order of magnitude grounds. These considerations are of fundamental importance. In the establishment of a mathematical model of running performance over distances from, say, 100 m to 10,000 m, corresponding to the extremes at which races are currently conducted in an athletic stadium, the following approach may justifiably be adopted. Attention may be concentrated on the energetics of the motion of the centre of mass of the runner, the small-scale cyclical variations of energy level associated with the stride pattern may be ignored with insignificant loss of accuracy, and the horizontal velocity of the runner's centre of mass may be regarded as a smooth, continuous function of time. These considerations underlie the discussion throughout the present paper.

Athletes adopt quite different approaches to sprint and long distance races, with middle distance events displaying intermediate features. Broadly, the characteristics are as follows. In a sprint event, an athlete exerts maximum available power throughout the course of the race. In distance events the runner, having accelerated from rest, laps at an essentially even pace, dictated by the length of the race. Short bursts of pace may be inserted from time to time, but any sustained increase in exertion is left to the final stages of the race. These differences between short and long races are largely explained by the important influence of lactic acid production on running performance. These mat-

ters will be considered further in section 3.3, where the kinetics of chemical energy conversion is considered, and in section 6.1, which contains a general discussion of race strategy in relation to the chemical processes involved in energy release.

2. SURVEY OF MATHEMATICAL MODELS OF RUNNING

2.1. *Methods based on the application of Newton's second law*

The first of a number of major contributions associated with Professor A. V. Hill and his co-workers was that of Furusawa *et al.* (1927a), who used Newton's second law of motion—rate of change of momentum is equal to the sum of the applied forces—to derive a relationship between running speed, v, distance traversed, x, and time from rest, t. Furusawa *et al.* (1927a) expressed the net applied force in terms of a driving or propelling force, F_D, from which was substracted an internal resistive force F_R. Thus the equation of motion of the centre of mass of the runner took the form

$$m \frac{d^2 x}{dt^2} = F_D - F_R. \tag{1}$$

The driving force F_D was taken by Furusawa *et al.* (1927) to be constant and was expressed as a fraction of body weight, whilst the resistive force F_R was ascribed to the 'viscosity' of the muscles, and was taken to be proportional to running speed.

Thus the equation of motion of the runner due to Furusawa *et al.* (1927a) could be expressed in the form

$$m \frac{d^2 x}{dt^2} = b \, mg - \frac{m v}{a} \tag{2}$$

where a and b are constants, m is mass, $v = dx/dt$, and g is acceleration due to gravity.

Integration of equation (2) yielded

$$x = abg \left[t - a \left(1 - \exp \left(-t/a \right) \right) \right]. \tag{3}$$

From equation (3) the velocity, v, at time t is given by

$$\frac{dx}{dt} = v = abg \left[1 - \exp \left(-t/a \right) \right]. \tag{4}$$

Equation (4) shows that for large values of t the velocity v asymptotically approaches a limiting value, which may be represented by V_{max}.

Thus, from equation (4)

$$V_{max} = abg. \tag{5}$$

Furusawa *et al.* (1927a) also reported the results of tests to establish the values of the constants a and b for different individuals, ranging from first-class sprinters, through good short distance runners and other athletes, and the tests also included non-athletes.

Further tests to establish the values of a and b were conducted by Best and Partridge (1929, 1930). In their 1930 paper the authors include the results of electrical timings on Percy Williams who, at the time, was the (joint) world record holder of the indoor 60 yard sprint. For Williams the values of $a = 1.34$ s and 1.35 s were established from two separate tests; in both cases a figure of $b = 0.79$ was deduced.

The paper by Best and Partridge (1929) included the results of further tests on athletes. Two types of experiment were conducted. The first set of tests were entirely conventional, involving the electrical timing of athletes running in the normal way. The second set of tests involved a new feature. In these tests the runner was subjected to a constant external resistance. From a belt fastened round the athlete's waist, a long chord was passed over a brake drum, mounted on a horizontal shaft. The resistance imposed on the runner could be adjusted. With this additional constraint, the results were shown to be consistent with the mathematical form of equations (3) and (4). Hence the authors felt entitled to claim 'that the internal viscous resistance of the muscles is not just a hypothesis invented to make the equation fit the observations, but is real, in the sense that it has identically the same effect as an external added resistance'.

Fenn (1930) was the first scientist to contest the physical arguments used to justify the 'viscous' resistance term in equation (2), and Hill (1938) later accepted that the case for using this reasoning could not be sustained. However, this type of analysis has by no means fallen into oblivion. If the doubtful physical basis of the analysis is ignored, equations (3) and (4) do provide a very reasonable mathematical fit to measured data for runners moving from rest over a distance of 50 or 60 m, when the appropriate values of a and b are chosen. Henry and Trafton (1951) have made further assessments of the magnitudes of the constants a and b by measuring the distance–time history of runners electronically.

Although Hill was well aware of the effects of air resistance (see Hill (1927)), he did not explicitly include them in the analysis he developed with Furusawa et al. (1927a). It is now known (see for example Ward-Smith, 1984) that Hill (1927), underestimated the influence of air resistance.

Keller (1973, 1974) analyzed both sprinting and distance events. Keller calculated that events run over distances less than 291 m could be classified as dashes or sprints, and involved a continuous acceleration from rest. For distances in excess of 291 m, Keller deduced that the final stages of the race should be run at constant velocity. In his analysis of these events, for the acceleration phase, Keller (1973, 1974) used an equation of motion identical in form to that of equation (2), due to Furusawa et al. (1927), although his notation was different.

Senator (1982) has added the effects of air resistance to equation (1) so that, using the present notation, the force balance can be expressed as

$$m\frac{d^2x}{dt^2} = F_D - F_R - D \qquad (6)$$

where D is the aerodynamic drag force.

In order to progress the analysis of equation (6), Senator (1982) combines the internal resistive force, F_R, and the external air resistance, D, by the empirical law

$$F_R + D = c\left(\frac{dx}{dt}\right)^n \qquad (7)$$

where c and n are constants.

There can be no physical justification for this arbitrary association of the terms, and so the subsequent analysis must be open to question.

On physical grounds it is preferable to regard the driving forces F_D and the internal resistive force, F_R, as combined into a single net driving force, say $F_D' (= F_D - F_R)$, as the free body diagram, Fig. 1 demonstrates.

Vaughan (1983a, b) has used this approach, writing Newton's law in the form

$$m\frac{dv}{dt} = m\frac{d^2x}{dt^2} = F_D' - D. \qquad (8)$$

Vaughan considers the empirical law

$$F_D' - D = m[A_1 - B_1 v^\alpha - C_1 v^2] \qquad (9)$$

where A_1, B_1, C_1 and α are assumed constant.

The aerodynamic drag D can be expressed in the form

$$D = \tfrac{1}{2}\rho v^2 S C_D \qquad (10)$$

where ρ is the density of the air, S is the frontal projected area of the runner and C_D is the drag coefficient. Comparing equations (9) and (10), the coefficient C_1 can be written as

$$C_1 = \frac{\rho S C_D}{2m}. \qquad (11)$$

Examination of equations (8) and (9) reveals that

$$F_D' = m(A_1 - B_1 v^\alpha). \qquad (12)$$

By measuring the performance of four University sprinters, Vaughan (1983a) obtained values for A_1, B_1 and α. The quantity α was evaluated as 0.7, A_1 ranged from 10.26 m s^{-2} to 10.65 m s^{-2}, whilst B_1, in units m$^{0.3}$ s$^{-1.3}$, ranged between 1.93 and 2.02.

Whilst Vaughan (1983a, b) has avoided the explicit use of an internal resistive force, the justification for

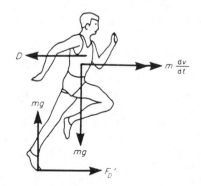

Fig. 1. Free body diagram of the forces acting on a runner.

340   A. J. WARD-SMITH

the empirical law, equation (12), for the net driving force F'_D has not been established by reference to the physical processes involved in running.

In summary, it must be concluded that available mathematical analyses of sprinting based on Newton's second law are unsatisfactory. All of the methods discussed suffer from at least two of the following deficiencies:

(1) The inclusion of empirical laws which have not been justified by reference to the physical processes involved in running.

(2) The velocity tends asymptotically to a constant maximum value with increase in t (a result which is inconsistent with actual practice).

(3) Because the velocity tends asymptotically to a maximum value the method cannot be conveniently extended to provide realistic estimates of running performance over long distances.

2.2. *Methods based on energy considerations*

An alternative approach to the formulation of a mathematical model of running is by reference to energy considerations. Furusawa et al. (1927b) recognised the importance of the energetics of running, but preferred to base their mathematical analysis of sprinting on Newton's laws of motion (Furusawa et al., 1927a).

The main initiative for the application of energy considerations rested with Lloyd, in a series of papers, (Lloyd, 1966, 1967; Lloyd and Moran, 1966). The basis of the method is the relationship that exists between the conversion of chemical energy and the related energy output of the runner. Lloyd (1967) proposed that the available chemical energy, C, consisted of two terms, one expressing the energy available from a store, the other term denoted the availability of energy at a steady rate. A separate expression for the energy spent by a runner in accelerating from rest, E_R, was derived by Lloyd (1967).

In principle a balance can be established between C and E_R. However, in order to proceed along these lines, Lloyd (1967) and Lloyd and Moran (1966) found it necessary to introduce factors for the mechanical efficiency of energy conversion, which were applied to certain of the terms comprising E_R. Using this approach Lloyd (1967) and Lloyd and Moran (1966) were able to correlate world running records, not only for sprinting events, but also for middle- and long-distance races. In this the authors were very successful. However, the correlation was achieved only by the incorporation of the mechanical efficiency factor; Lloyd (1967) proposed a figure of approximately 25%, whereas Lloyd and Moran (1966) suggested a value nearer 50%.

The introduction of the mechanical efficiency factor is unnecessary and obscures the precise nature of the overall energy balance in running.

In section 3 a new analysis of running is proposed. It is intended to use an approach similar to that of Lloyd (1967) and Lloyd and Moran (1966), but to improve the physical base upon which the mathematical model depends, by incorporating an improved formulation of the kinetics of chemical energy conversion and by avoiding the use of the mechanical efficiency factor.

3. A NEW ANALYSIS OF THE ENERGETICS OF RUNNING

3.1. *Basic energy and power equations*

In this section a new mathematical model of the energetics of running is derived. To start with the processes of energy exchange occurring during running are considered. Figure 2, based on Ward-Smith (1984), is a diagramatic representation of these processes. Although the energy released by chemical conversion passes through a number of intermediate stages, a clear overall energy balance can be formulated, as the discussion in the Introduction has already indicated.

A thermodynamic overview indicates that most of the chemical energy is ultimately transformed into thermal energy, which from a formal standpoint is manifested as an increase of the internal energy of the body. Any imbalance in the transition of chemical energy into mechanical energy is accompanied by an enthalpy rise, which is immediately reflected as a change in the internal energy of the system. However, the other reason for the increase in the internal energy is the irreversible degradation of mechanical energy into thermal energy.

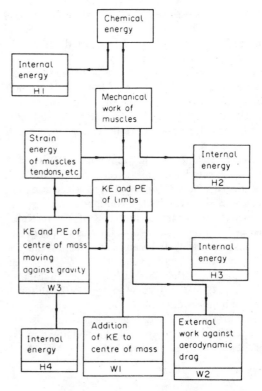

Fig. 2. Representation of the energy exchanges occurring during running (after Ward-Smith, 1984).

Applying the principles of the first law of thermodynamics, it follows that the chemical energy released, C, is equal to the sum of the useful work done by the centre of mass of the runner, W, plus the mechanical energy degraded into thermal energy, H.

$$C = H + W \qquad (13)$$

where using the notation of Fig. 2

$$H = H_1 + H_2 + H_3 + H_4$$

and

$$W = W_1 + W_2 + W_3.$$

Averaged over a number of strides, $W_3 = 0$, if the small change in potential energy of the centre of mass of a runner between the start and finish of a race on level ground is ignored.

Differentiating equation (13) with respect to time t, there results

$$\frac{dC}{dt} = \frac{dH}{dt} + \frac{dW}{dt}$$

which can be written in the form

$$P = P_H + P_e \qquad (14)$$

where $P = dC/dt$, $P_H = dH/dt$ and $P_e = dW/dt$, and the symbol P (with or without suffix) denotes power.

3.2. *The power required for running*

The individual contributions to the power equation (14) can be identified as follows.

The rate at which kinetic energy is added to the centre of the mass of the runner is given by

$$\frac{dW_1}{dt} = m v \frac{dv}{dt}.$$

The rate of working against aerodynamic drag is (in the absence of any wind)

$$\frac{dW_2}{dt} = Dv = \tfrac{1}{2}\rho v^3 S C_D$$

where ρ is the air density, S is the frontal projected area of the runner and C_D is the drag coefficient.

The rate of increase of thermal energy is considered in depth by Ward-Smith (1984). Tests on subjects running on a treadmill, as reported by Davies (1980) and Margaria (1976), can be represented by the equation

$$\frac{dH}{dt} = Av.$$

Additionally, Ward-Smith (1984) demonstrated that the relation can be extrapolated to apply to sprinting speeds. Thus the power required for running is given by summing the foregoing contributions and is

$$P_H + P_e = Av + \tfrac{1}{2}\rho v^3 S C_D + mv \frac{dv}{dt}. \qquad (15)$$

3.3. *The kinetics of chemical energy conversion*

The fundamental source of chemical energy is the exergonic reaction during which adenosine triphos-phate (ATP) splits into adenosine diphosphate (ADP) and phosphoric acid. Since muscle contains only a very small amount of ATP, it has to be continuously resynthesised, at the same rate at which it is split. The energy for this is provided by the cleavage of creatine phosphate into creatine and phosphoric acid. Since the two reactions are in series and their energy contents are roughly the same, it is convenient to introduce the term phosphagen to describe all the substances containing high energy phosphate. The cleavage of phosphagen can therefore be regarded as the primary energy source in muscular activity. For extended muscular exercise phosphagen must be resynthesised continuously utilising the energy from food combustion and/or glycolysis.

For prolonged exercise the oxidative mechanism prevails, and there is a maximum rate of energy release, represented by R, that can be maintained by this mechanism. Tests conducted from rest at a constant rate of energy expenditure have shown (see, e.g. Margaria, 1976) that the aerobic power, P_{aer}, grows exponentially and hence it may be described by an equation of the form

$$P_{aer} = R(1 - \exp(-\lambda t))$$

which, upon integration, yields the amount of energy contributed by the oxidative mechanism as

$$C_{aer} = R t - \frac{R}{\lambda}(1 - \exp(-\lambda t)). \qquad (16)$$

During the early stages of exercise, and under conditions of supramaximal exercise, the energy available from the aerobic mechanism is insufficient and it is supplemented by energy from the anaerobic mechanism. This energy source is vital in sprint events. Even during less vigorous exercise, as in long-distance running, it is an energy resource which can be usefully exploited to augment the supply from the aerobic mechanism. Initially, the anaerobic power is provided entirely by phosphagen cleavage, but when the phosphagen is depleted to about one half its initial level, the glycolytic mechanism, characterised by the production of lactic acid, is activated.

The production of lactic acid is associated with considerable physical discomfort and quickly leads to diminished levels of performance. It is for this reason that athletes in middle and long distance races avoid excessive exertion in the early stages of the race, as, in this way, activation of the glycolytic mechanism is delayed to the final stages of the race.

Denoting the capacity of the alactic mechanism by E_a and that of the glycolytic mechanism by E_g, the total capacity of the anaerobic mechanism, E_o, is given by

$$E_o = E_a + E_g.$$

In the longer distance races (i.e. from 1500 m upwards) performance is determined by overall energy balances, including the contribution from E_o. As will become evident, provided the premature production of lactic acid is avoided, the time history of the release of energy by the anaerobic mechanism is of no significance in determining the relation between overall running time and race distance. In sprint events, and to a lesser extent in the middle distance events, the rate of release of energy from the anaerobic mechanism has an important influence on running performance, and in order to predict running performance in these events a mathematical expression is required for the variation with time of the maximum anaerobic power available to the athlete. This requirement may be approached in the following way. First, it is appropriate to draw attention to the following conditions (Margaria, 1976), which the relationship must satisfy: (1) the total amount of energy available from the cleavage of phosphagen plus that from the glycolytic mechanism is finite (and equal to E_o); (2) the available anaerobic power is a maximum at the initiation of movement ($t = 0$); and (3) the anaerobic power available reduces with increase in t. In order to incorporate these conditions into a mathematical relation we shall adopt the hypothesis due to Margaria (1976), namely that, prior to the activation of the glycolytic mechanism, the rate of energy release due to oxidation is dependent upon the amount of split phosphagen.

The total amount of energy from the combined anaerobic sources which is unused and potentially available at time t, will be denoted by E. Hence, following the hypothesis of Margaria (1976), we write

$$P_{aer} = k(E_o - E)$$

where k is a constant of proportionality. This expression may be written, on substituting for P_{aer}, as

$$E_o - E = \frac{R}{k}(1 - \exp(-\lambda t))$$

which, upon differentiation, yields

$$-\frac{dE}{dt} = P_{an} = \frac{R\lambda}{k}\exp(-\lambda t).$$

Replacing $R\lambda/k$ bt P_{max}, the maximum power available from the alactic mechanism, the final expression for P_{an} is obtained. Thus

$$P_{an} = P_{max}\exp(-\lambda t) \qquad (17)$$

an expression which satisfies all of the conditions previously listed.

Strictly, the use of equation (17) to describe the maximum anaerobic power available from rest can only be justified prior to the activation of the glycolytic mechanism. However, in the absence of any obvious alternative description, it will be employed here to describe the entire range of the anaerobic mechanism. In doing so, the following observations are noted: (1) there is no observed sharp change to running performance corresponding to the activation of the glycolytic mechanism (as distinct from the later deterioration corresponding to the accumulation of lactic acid); (2) equation (17) has the correct anaerobic capacity for the combined energy sources of phosphagen cleavage and glycolysis, and (3) it is known

(Margaria, 1976) that the maximum available power of the glycolytic mechanism is much lower than P_{max}. In all these three respects the use of equation (17) is consistent with or does not drastically violate observed running performance or results of physiological tests. Consequently equation (17) will be used to represent the maximum available anaerobic power for all t.

Integration of equation (17) yields

$$C_{an} = E_o - E = \frac{P_{max}}{\lambda}(1 - \exp(-\lambda t)). \qquad (18)$$

From equation (18) with $t = \infty$, the relation

$$E_o = \frac{P_{max}}{\lambda}$$

is derived, thereby relating the capacity of the anaerobic mechanism to the maximum anaerobic power. It will subsequently be shown in section 6.3 that the available data for running performance are in reasonable agreement with this expression.

A knowledge of the maximum rate at which energy can be extracted from the anaerobic mechanism is not only important to the calculation of sprinting performance, but is also important to the understanding of the strategy employed by athletes in races over long distances. For this reason the power and energy equations will now be set down for an athlete operating at maximum available power from rest. In section 6.1 a full discussion will be presented of the interpretation of calculations based on these equations.

By summing the contributions from the aerobic and anaerobic sources the equation for the power available from chemical energy conversion is obtained. It is

$$P = P_{an} + P_{aer} = (P_{max} - R)\exp(-\lambda t) + R. \qquad (19)$$

The corresponding energy equation is

$$C = C_{an} + C_{aer} = \left(\frac{P_{max} - R}{\lambda}\right)(1 - \exp(-\lambda t)) + Rt. \qquad (20)$$

Equations (19) and (20) can be written in the alternative forms

$$P = P_{max}\exp(-\lambda t) + R(1 - \exp(-\lambda t)) \qquad (21)$$

and

$$C = E_o[1 - \exp(-\lambda t)] + R\left(t - \frac{1}{\lambda}(1 - \exp(-\lambda t))\right) \qquad (22)$$

where the first and second terms on the right-hand side of equations (21) and (22) represent the contributions from the anaerobic and aerobic mechanisms, respectively. These formulations of the power and energy equations clearly demonstrate the dominant influence of the anaerobic sources during the initial stages of energy conversion and the increasing influence of the aerobic contribution with the passage of time.

The physical arguments underlying the mathematical model derived by Lloyd (1966, 1967) to describe the processes of chemical energy conversion are less detailed and less plausible than those employed here. However, equations (19) and (20) are mathematically identical in form to the equations for the power and energy due to chemical conversion proposed by Lloyd. His equations can be recovered by introducing the parameter S_o defined by

$$S_o = \frac{P_{max} - R}{\lambda} = \frac{P_{max} - R}{P_{max}} E_o. \quad (23)$$

It should be noted that E_o, but not S_o, is a measure of the oxygen debt sustained during a period of running. The parameter S_o, although possessing the units of energy, should be regarded as a variable introduced for mathematical convenience, and having no special physical significance.

3.4. *Final forms of the power and energy equations of a runner*

Substitution of equations (15), (19) and (23) in equation (14) yields the power equation

$$\lambda S_o \exp(-\lambda t) + R = Av + \tfrac{1}{2}\rho v^3 S C_D + mv\frac{dv}{dt}. \quad (24)$$

The corresponding energy equation is derived by integration of equation (24) with respect to t, and substitution of the appropriate initial conditions. This leads to the energy equation

$$S_o(1 - \exp(-\lambda t)) + Rt = Ax + \tfrac{1}{2}\rho S C_D$$
$$\times \int_0^t v^3\, dt + \tfrac{1}{2}mv^2 \quad (25)$$

where

$$x = \int_o^t v\, dt. \quad (26)$$

It is to be observed that in formulating the power and energy equations (24) and (25), full account is taken of all terms contributing to the balances and, in contrast to Lloyd's formulations, there is no necessity to introduce the concept of mechanical efficiency.

Equations (24) to (26) do not have an exact solution. However, they can be successfully approximated or can be rewritten in such a way that they are amenable to numerical integration. In formulating the equations for these purposes we shall also take into account the influence of body mass on running performance.

On dimensional grounds, it can be argued that the parameters S_o, R and A are directly proportional to body mass, m. On the other hand S is proportional to $m^{2/3}$. Dividing through equation (25) by m, it can be rewritten as

$$S_o^*(1 - \exp(-\lambda t)) + R^* t = A^* x$$
$$+ K^* \int_o^t v^3\, dt + \frac{v^2}{2} \quad (27)$$

where

$$S_o^* = \frac{S_o}{m}, \quad R^* = \frac{R}{m}, \quad A^* = \frac{A}{m} \quad (28)$$

and

$$K^* = \frac{\rho S C_D}{2m}. \quad (29)$$

4. APPLICATION OF THE ANALYSIS TO RESULTS FROM THE OLYMPIC GAMES

Athletic records are an extremely important source of experimental data obtained under carefully controlled conditions. From time to time records are broken, testifying to the improved levels of performance which are attained. In a full evaluation of experimental data it is possible to take account of these long-term improvements in performance (see, e.g. Ward-Smith, 1984). However, in the present context, these changes are of insufficient importance to be considered explicitly. Instead representative magnitudes of the biophysical parameters will be calculated, based on the average performance of athletes in the Olympic Games over the years from 1960 to 1976.

It is convenient to consider initially the magnitudes of those parameters amongst S_o^*, R^*, A^* and K^* which may be quantified most easily.

We start with the parameter K^*, where

$$K^* = \frac{\rho S C_D}{2m}.$$

For normal conditions of temperature and pressure, the atmospheric density is given by $\rho = 1.22 \text{ kg m}^{-3}$. Using the data of Davies (1980) a representative value for the product $S C_D$ of a runner is 0.385 m^2. Assuming a nominal mass of 70 kg, K^* is evaluated as

$$K^* = \frac{1.22 \times 0.385}{2 \times 70} = 3.355 \times 10^{-3} \text{ m}^{-1}.$$

The data of Margaria (1976) and Davies (1980), based on steady-state running experiments on a treadmill, and the extrapolation of these results to sprinting speeds by Ward-Smith (1984), are all in close accord with a value of

$$A^* = 3.9 \text{ J m}^{-1} \text{ kg}^{-1}.$$

For the most part, long distance events are run at or close to a constant speed V_{av}, and the capacity of the anaerobic mechanism is almost entirely depleted, so that the running performance is not directly affected by λ. Under such circumstances the time T to run a distance s is obtained by approximating equation (27) by the expression

$$S_o^* + R^* T = A^* s + K^* V_{av}^3 T + \frac{V_{av}^2}{2} \quad (30)$$

where

$$V_{av} = \frac{s}{T}. \quad (31)$$

Equation (30) can be recast in the form

$$s' = \frac{S_o^*}{A^*} + \frac{R^*}{A^*} T \quad (32)$$

344 A. J. WARD-SMITH

where

$$s' = s + \frac{V_{av}^2}{A^*}[K^*s + 0.5].\qquad(33)$$

The second term is of second order.

Thus a plot of the time taken to run long-distance events (say 1500 m up to 10,000 m) should yield a linear relationship between s' against T. The intercept is a measure of the magnitude of S_o^* and the slope is directly related to R^*.

The results of events in recent Olympic Games have been collected by McWhirter and McWhirter (1980). Table 1 lists the average finishing times of the first three runners in the 1500 m, 5000 m and 10,000 m races in the Games of 1960, 1964, 1972 and 1976. The results of the 1968 Olympics, held in Mexico City, were appreciably affected by the reduced oxygen content of the atmosphere at altitude, and have been excluded from the analysis. The corresponding values of s', computed using equation (32) with $K^* = 3.355 \times 10^{-3}$ m and $A^* = 3.9$ W s m^{-1} kg^{-1}, are given in Table 1. The three data points of Table 1 are plotted in Fig. 3 and are in excellent agreement with the predicted linear relationship, yielding a value for R^* of 23.2 W kg^{-1}.

An approximate value of S_o^* can be determined from Fig. 3, but a more accurate approach is to consider data for a wider range of events, including those over shorter distances. The quantity λ can be evaluated at the same time.

Substituting the values $x = s$, $t = T$, and $v = V_{av}$, equation (27) can be approximated by the expression

$$S_o^*(1 - \exp(-\lambda T)) + R^*T = A^*s + K^*V_{av}^3T + \frac{V_{av}^2}{2}\qquad(34)$$

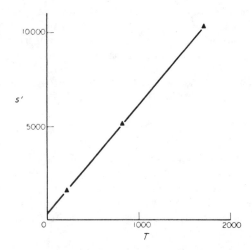

Fig. 3. Variation of s' (m) with T (s), based on the average time of the first three runners in the 1500 m, 5000 m and 10,000 m events at the Olympic Games of 1960, 1964, 1972 and 1976.

where the approximation

$$K^*V_{av}^3T = K^* \int_o^T v^3 \, dt$$

has been introduced to simplify the expression.

The energy balance represented by equation (34) is considered in Table 2, where the average finishing times of the first three runners in the Olympic Games of 1960, 1964, 1972 and 1976 are analysed.

In evaluating the right-hand side of equation (34) the quantities $A^* = 3.9$ J kg^{-1} m^{-1} and $K^* = 3.355 \times 10^{-3}$ m^{-1}, as previously determined, can be substituted directly. The magnitudes of S_o^*, λ and R^* were evaluated such that the maximum discrepancy in the energy balance represented by equation (34) was minimised, as indicated by the values in the right-hand column of Table 2. Using this criterion it was found that the data of Table 2 were best correlated taking

$$S_o^* = 900 \text{ J kg}^{-1}, \quad \lambda = 0.03 \text{ s}^{-1}$$
$$\text{and} \quad R^* = 23.5 \text{ W kg}^{-1}.$$

The value of R^* used here supersedes the magnitude determined from Fig. 2, as it provides a rather better correlation over the full range of distances considered

Table 1. Average finishing time of the first three runners in the Olympic Games of 1960, 1964, 1972, 1976. These data are plotted in Fig. 3.

s (m)	T (s)	V_{av} (m s^{-1})	s' (m)
1500	218.24	6.8732	1567
5000	816.48	6.12385	5166
10000	1686.83	5.9283	10307

Table 2. Calculation of energy balances, based on the average time of the first three runners for all of the events from 100 m to 10,000 m in the Olympic Games of 1960, 1964, 1972, 1976. The quantities C^*, H^* and W^* are evaluated, using the data of Table 3, from the relations

$$C^* = S_o^*(1 - \exp(-\lambda T)) + R^*T$$
$$H^* + W^* = A^*s + K^*sV_{av}^2 + V_{av}^2/2$$

s (m)	T (s)	V_{av} (m s^{-1})	C^* (J kg^{-1})	$H^* + W^*$ (J kg^{-1})	$C^* - H^* - W^*$ (J kg^{-1})	$(C^* - H^* - W^*)/C^*$
100	10.19	9.813(5)	476.5	470.4	+6.1	+1.3%
200	20.40	9.804	891.4	892.6	-1.2	-0.1%
400	44.93	8.903	1722	1706	+16	+0.9%
00	105.48	7.584	3341	3303	+38	+1.1%
0	218.24	6.873	6027	6111	-84	-1.4%
	816.48	6.124	20088	20148	-60	-0.3%
	1686.83	5.928	40541	40197	+344	+0.8%

in Table 2. That the correlation of data is very successful is indicated by reference to Table 2. Events in the full range from 100 m to 10,000 m are considered, yet the maximum discrepancy in the energy balance nowhere exceeds $\pm 1.4\%$.

A summary of the values of the parameters S_o^*, λ, R^*, K^* and A^*, on which the correlation of Table 2 is based, is to be found in Table 3.

An alternative presentation of Table 2 is as follows.

Equation (34) can be rewritten, by substituting for V_{av} using equation (31), in the form

$$S_o^*(1 - \exp(-\lambda T)) + R^*T = A^*s + \frac{K^*s^3}{T^2} + \frac{s^2}{2T^2}$$

(35)

which can be written, after rearrangement, as

$$2S_o^*T^2(1 - \exp(-\lambda T)) + 2R^*T^3 = 2A^*sT^2 + 2K^*s^3 + s^2 \quad (36)$$

which is an implicit relation between s and T. The solution of equation (36), using the coefficients listed in Table 3 is presented in Table 4. Column 2 lists the measured times, whilst column 3 lists the solution of equation (36).

5. VELOCITY–TIME HISTORY IN SPRINTING

It was possible to pursue the analysis considered in section 4 without detailed reference to the relationship between speed and distance covered during the course of a race. Distance events, in particular, are governed

Table 3. Magnitude of the parameters used to correlate the data of Table 2

Variable	Magnitude
S_o^*	900 J kg^{-1}
λ	0.03 s^{-1}
R^*	23.5 W kg^{-1}
A^*	3.9 J m^{-1} kg^{-1}
K^*	3.355×10^{-3} m^{-1}

Table 4. Comparison of prediction based on equation (36) with Olympic games results

1	2	3	4	5
s (m)	T (s)	T (s)	ΔT (s)	$\dfrac{\Delta T}{T}$ (%)
100	10.19	10.09	+0.10	+1.0
200	20.40	20.42	−0.02	−0.1
400	44.93	44.50	+0.43	+1.0
800	105.48	104.15	+1.33	+1.3
1500	218.24	221.49	−3.25	−1.5
5000	816.48	818.89	−2.41	−0.3
10000	1686.83	1673.04	+13.79	+0.8

Column 2. Average time of the first three finishers in the 1960, 1964, 1972, 1976 games.
Column 3. Equation (36) evaluated using the parameters of Table 3.

by the overall energy considerations discussed in section 4.

However the analysis there made use of the approximation

$$K^* \int_o^t v^3 \, dt = K^* V_{av}^3 T$$

and this simplification is not very realistic in sprint events.

Using equations (28) and (29), equation (24) can be rewritten as

$$\frac{dv}{dt} = \frac{1}{v}[\lambda S_o^* \exp(-\lambda t) + R^* - A^*v - K^*v^3]. \quad (37)$$

Equation (26) can be written in the form

$$\frac{dx}{dt} = v. \quad (38)$$

Equations (37) and (38) are a pair of ordinary differential equations. A computer programme was written and these equations were integrated to obtain $v = v(t)$ and $x = x(t)$ using a numerical scheme based on the fourth order Runge-Kutta method. The application of this method is widely discussed in the literature, see for example Chow (1979).

There is one detailed point that must be taken into account in solving equation (37). A singularity exists at the condition $t = 0$, indicating that $dv/dt = \infty$; in other words the mathematical model indicates (unrealistically) that the runner experiences an infinite acceleration as he moves from rest. As soon as the velocity is finite, so a finite value of dv/dt is indicated by the analysis.

In the computer programme, the singularity at $t = 0$ is treated in the following way. The velocity of the runner a small increment of time Δt from rest is calculated by rewriting equation (37) as

$$v \, dv = [\lambda S_o^* + R^*] \, dt \quad (39)$$

which integrates to yield

$$v = \sqrt{2}[\lambda S_o^* + R^*]^{1/2} (\Delta t)^{1/2} \quad (40)$$

during which time the runner travels a distance

$$x = \sqrt{\tfrac{8}{9}}[\lambda S_o^* + R^*]^{1/2} (\Delta t)^{3/2}. \quad (41)$$

Once conditions for the first increment in time Δt had been established using equations (40) and (41), the remaining steps in the numerical integration proceeded by solving equations (37) and (38).

The results of applying the above procedure to a 100 m race, using the variables listed in Table 3, are shown in Fig. 4. A time-step of 0.05 s was used in the numerical integration. The value of $T = 10.23$ s computed in this way compares well with the Olympic average time of 10.19 s, and represents an improvement over the value of $T = 10.09$ s, evaluated using equation (36).

348   A. J. WARD-SMITH

Table 6. Magnitudes of some biophysical parameters of
world-class athletes

Variable	Units	Margaria (1976)	Present calculations
P^*_{max}	W kg^{-1}	56	50.5
R^*	W kg^{-1}	25	23.5
E^*_o	J Kg^{-1}	1900	1700
S^*_o	J kg^{-1}	1000	900
λ	s^{-1}	$\begin{cases} 0.03 \\ 0.023 \end{cases}$	0.03

Column 3. Approximate values derived from Margaria
(1976).
Column 4. Data derived from the present correlation.

data quoted by Margaria (1976). Using the fact that
1 cal = 4.18 J, a value of 48 kcal kg^{-1} h^{-1}, quoted on
p. 34 of Margaria (1976) converts to P^*_{max}
$\simeq 56$ W kg^{-1}, where the starred notation indicates
that a quantity is evaluated per unit mass of body
weight. The figure for E^*_o on p. 34, obtained by
summing the lactacid and alactacid contributions, of
450 cal kg^{-1} corresponds with $E^*_o \simeq 1900$ J kg^{-1}.
These data for P^*_{max} and E^*_o were average values
obtained from a range of physiological tests on young
normal subjects. Taking 1 ml of O_2 as equivalent to
5 cal, a value of $V^{max}_{O_2}$ of 73 ml kg^{-1} min^{-1}, which is the
average of three figures deduced by Margaria (1976),
p. 44, from the world records over 1500 m, 5000 m and
10,000 m which existed in 1969, yields R^*
$\simeq 25$ W kg^{-1}. Using the above data, from equation
(23) the quantities $S^*_o = 1000$ J kg^{-1} and $\lambda = 0.03$ s^{-1}
are evaluated. From separate physiological tests,
(Margaria, 1976) shows that the half-life of the ex-
ponential law describing the aerobic mechanism is
about 30 s, yielding an alternative value for λ of
0.023 s^{-1}. Bearing in mind the very different
backgrounds from which the data of Table 6 were
derived, the agreement between the two sets is seen to
be extremely good.

7. CONCLUDING REMARKS

1. A survey has been made of existing mathematical
models of running performance. It is shown that
methods based on the application of Newton's second
law of motion have physical limitations.

2. A method of analysis based on the first law of
thermodynamics is developed. The method represents
an extension of ideas originally proposed by Lloyd
(1967). Unlike Lloyd (1967), the present analysis
avoids the use of mechanical efficiency factors, and
properly accounts for each term in the energy balance.
An improved description of the kinetics of chemical
energy conversion has been incorporated in the
analysis.

3. A mathematical expression (equation 36) is derived which provides a relationship, between race
distance and the time taken by athletes of world-class

to run that distance. The form of equation (36)
provides an excellent correlation of the data, for
distances of 100 m up to 10,000 m.

4. The velocity–time relationship for sprinting
100 m at maximum available power has been established by numerical integration of the power
equation. The curve indicates that the peak velocity is
achieved in the middle stages of the race, a result which
is consistent with practice, and one which earlier
theories have failed to predict.

5. The results obtained in the present paper have
application in the calculation of long-jump performance and, with suitable modification, can also be used
to predict the effects of the wind on sprinting
performance.

Acknowledgement—I am grateful to Professor A. J. Reynolds
and Mr. D. W. Murrie for helpful discussions and for
commenting on an earlier draft of this paper.

REFERENCES

Best, C. H. and Partridge, R. C. (1929) The equation of motion
of a runner, exerting a maximal effort. *Proc. R. Soc.* B 103,
218–225.
Best, C. H. and Partridge, R. C. (1930) Observations on
Olympic athletes. *Proc. R. Soc.* B 105, 323–332.
Chow, C.-Y. (1979) *An Introduction to Computational Fluid
Mechanics.* John Wiley, New York.
Dagg, A. I. (1977) *Running, Walking and Jumping. The Science
of Locomotion.* Wykeham Publications, London.
Davies, C. T. M. (1980) Effect of wind assistance and
resistance on the forward motion of a runner. *J. appl.
Physiol.* 48, 702–709.
Fenn, W. O. (1930) Frictional and kinetic factors in the work
of sprint runners. *Am. J. Physiol.* 92, 583–610.
Furusawa, K., Hill, A. V. and Parkinson, J. L. (1927a) The
dynamics of sprint running. *Proc. R. Soc.* B 102, 29–42.
Furusawa, K., Hill, A. V. and Parkinson, J. L. (1927b) The
energy used in sprint running. *Proc. R. Soc.* B 102, 43–50.
Henry, F. M. and Trafton, I. R. (1951) Velocity curve of sprint
running. *Res. Q.* 22, 409–422.
Hill, A. V. (1927) The air resistance to a runner. *Proc. R. Soc.* B
102, 380–385.
Hill, A. V. (1938) The heat of shortening and the dynamic
constants of muscles. *Proc. R. Soc.* B 126, 136–195.
Ikai, M. (1967) Biomechanics of sprint running with respect
to the speed curve. *Medicine and Sport*, Part 2, pp. 282–290.
Karger, Basel/New York.
Keller, J. B. (1973) A theory of competitive running. *Phys.
to-day* 26, 43–47.
Keller, J. B. (1974) Optimal velocity in a race. *Am. math. Mon.*
81, 474–480.
Kellett, D. W., Mahon, M. and Willan, P. L. T. (1983) A
comparison of some biophysical characteristics in British
male sprinters and marathon runners. *J. Sports Sci.* 1,
76–77.
Lloyd, B. B. (1966) Energetics of running: An analysis of
world records. *Adv. Sci.* 22, 515–530.
Lloyd, B. B. (1967) World running records as maximal
performances. Oxygen debt and other limiting factors.
Circulation Res. XX, XXI, Supplement 1, 218–226.
Lloyd, B. B. and Moran, P. T. (1966) Analogue computer
simulation of the equation of motion of a runner. *J.
Physiol.* 186, 18–20.
Margaria, R. (1976) *Biomechanics and Energetics of Muscular
Exercise.* Clarendon Press, Oxford.
McWhirter, N. and McWhirter, R. (1980). *The Guinness Book
of Olympic Records.* Penguin Books, London.

A mathematical theory of running 349

Senator, M. (1982) Extending the theory of dash running. *J. biomech. Engng* **104**, 209-213.

Vaughan, C. L. (1983a) Simulation of a sprinter. Part 1. Development of a model. *Int. J. bio-med. Comput.* **14**, 65-74.

Vaughan, C. L. (1983b) Simulation of a sprinter. Part II.

Implementation on a programmable calculator. *Int. J. Bio-med. Comput.* **14**, 75-83.

Ward-Smith, A. J. (1984) Air resistance and its influence on the biomechanics and energetics of sprinting at sea level and at altitude. *J. Biomechanics* **17**, 339-347.